OSTEOBIOGRAPHIES

OSTEOBIOGRAPHIES
The Discovery, Interpretation and Repatriation of Human Remains

SUSAN PFEIFFER
Department of Anthropology, University of Toronto, Toronto, ON, Canada

ELSEVIER
ACADEMIC PRESS
An imprint of Elsevier

Academic Press is an imprint of Elsevier
125 London Wall, London EC2Y 5AS, United Kingdom
525 B Street, Suite 1650, San Diego, CA 92101, United States
50 Hampshire Street, 5th Floor, Cambridge, MA 02139, United States
The Boulevard, Langford Lane, Kidlington, Oxford OX5 1GB, United Kingdom

Notices
Knowledge and best practice in this field are constantly changing. As new research and experience broaden our understanding, changes in research methods, professional practices, or medical treatment may become necessary.

Practitioners and researchers must always rely on their own experience and knowledge in evaluating and using any information, methods, compounds, or experiments described herein. In using such information or methods they should be mindful of their own safety and the safety of others, including parties for whom they have a professional responsibility.

To the fullest extent of the law, neither the Publisher nor the authors, contributors, or editors, assume any liability for any injury and/or damage to persons or property as a matter of products liability, negligence or otherwise, or from any use or operation of any methods, products, instructions, or ideas contained in the material herein.

Library of Congress Cataloging-in-Publication Data
A catalog record for this book is available from the Library of Congress

British Library Cataloguing-in-Publication Data
A catalogue record for this book is available from the British Library

ISBN 978-0-12-823880-6

For information on all Academic Press publications
visit our website at https://www.elsevier.com/books-and-journals

Publisher: Nikki Levy
Acquisitions Editor: Anna Valutkevich
Editorial Project Manager: Sara Pianavilla
Production Project Manager: Maria Bernard
Cover Designer: Christian Bilbow

Typeset by STRAIVE, India

Contents

Acknowledgments *ix*
Introduction *xi*

1. Osteobiography: A way to see past lives **1**
What is a bioarchaeologist? 1
What is an osteobiography? 2
The tools of the trade 9
Terms, labels and designations 17
References 19

2. Rest in peace **21**
Prospect Hill 21
Snake Hill 30
Historic graves in broader context 35
References 36

3. Framework for Osteobiographies: Lower Great Lakes **37**
Prologue 37
History's story line 38
Archaeology's story line 41
The contemporary context 43
Summary 48
References 49

4 Mobile people making a go of it **51**
Prologue 51
So little to go on 53
The Hind Site 54
A child's distinctive skeleton 56
Looking back at the Hind Site 61
References 62

5. We enter the afterlife together **65**
Prologue 65
Background to the story 66
The excavation of ossuaries 68
Ossuary osteobiographies 69
Evidence of a disease that wasn't supposed to be there 71
Ossuary communities: Collective osteobiographies 74
Concluding remarks 80
References 80

6. Collaboration and repatriation **83**
Prologue 83
Construction of a collaborative framework 83
Preliminary steps 84
Meanwhile, other developments 86
Preparing for repatriation 88
Repatriation and after 94
Things to not do 96
References 97

7. Framework for osteobiographies: Eastern and Southern Africa **99**
Prologue 99
The lure of Africa 101
Modern humans: What role for Africa? 104
Osteobiographies of past Africans 109
References 110

8. Lake Turkana: An intriguing osteobiography **111**
Prologue 111
The Lake Turkana Basin 112
The pillar sites 114
The still incomplete picture 122
References 124

9. Southern African foragers: Geographical and historical context **125**
Prologue 125
The remarkable shrinking forager 131
A large number of small people 140
References 142

10. Southern African foragers: Life histories of past communities **145**
Approaching the question of body size 145
Summary 155
References 156

11. Interpersonal violence among Southern African foragers **161**
Prologue 161
Interpretive approaches to violent acts in the distant past 165
The skeletal evidence 166
What to make of it all? 170
Bearing witness 173
References 174

12. Repatriation in an African context **177**
Prologue 177
Unnamed persons 178
Named persons 183
References 191

13. Bones and objects of contention **193**
Prologue 193
The world of objects 194
The world of bones 196
Legacy collections 199
Beyond museums 205
The way forward 206
References 208

Index *211*

Acknowledgments

Many of the examples described here come from research that reflects not just my own efforts, but also that of an array of wonderful colleagues. Some of their names appear as coauthors on publications cited in the chapters, but there have been many others as well. I have been privileged to work with consulting archaeologists, museum curators, Indigenous spokespersons, anthropologists, students, and researchers from many branches of the anthropological, biological, and physical sciences. In both Canada and Africa, they have been openhearted, trustworthy guides and collaborators. I especially want to thank my many South African friends who helped orient me to a whole new world. It was perhaps foolhardy to plunge deeply into a new program of research in midcareer, but the patience and generosity of the University of Cape Town community and their colleagues across southern Africa buoyed my jet-lagged spirits, year after year.

Many people helped me with the creation of this book. Foremost in my thanks are the close colleagues and frequent collaborators who read and gave advice on whole sections. Ronald F. Williamson gave me feedback on the chapters about Canada, and Judith C. Sealy gave me feedback on the chapters about Africa. My draft text benefitted from the input of Devin Ward, who provided a learner's perspective. The maps unique to this book were produced by Andrew Stewart of Strata Consulting, Toronto. I thank all the people who responded to my requests for information or aid in areas of their own expertise. They include Amy Beresheim, Michael Chazan, Dan D'Agostino, Pierre de Maret, Kate Dougherty, Hilary Duke, Brendan Edwards, Victoria Gibbon, Robert Gordon, Kate Grillo, Lesley Harrington, Lisa Hildebrand, Richard Lee, Janet Monge, Shiona Moodley, Maria J. Odonovan, Sven Ouzman, Sam Schiller, Wilhelmina Seconna, Jeremy Silvester, Thivviya Vairamuthu, and Erin Wilson. I thank my husband, Eaton Lattman, for his sustained support, encouragement, and stylistic feedback. Despite the involvement of all these wonderful people, I take sole responsibility for the final product, including its views and opinions.

Introduction

This book explores what we can learn from the bones and teeth of people who died long ago. Its focus is on people of the past. They are our global antecedents from hundreds or thousands of years ago. Their many, diverse lifeways may be well documented or may be known only imperfectly. In concert with their archaeological context, their bones and teeth provide evidence of their lives. While this book introduces many of the research methods used, it is not a training manual. Rather, it is an exploration of how the scientific study of human remains can advance our understanding of the past. Each osteobiography is constructed within the framework of bioarchaeology, the scientific study of human remains within their archaeological context. Readers who are interested to know more about the field can use the citations herein to access journals and books related to bioarchaeology.

I chose case studies for this book that illustrate various dimensions of the relationships between people and their biosocial environments. The case studies show how information from archaeologically derived skeletons can correct past inaccuracies and expand our knowledge of the past. This is especially important for Indigenous communities whose histories, frequently written by others, are brief and often biased. Osteobiographies can enhance our understanding of past human struggles and resilience. The acts of discovery, study, and respectful treatment of human skeletons from the past are important to scholarship. They also affirm social values.

Methods for the excavation and study of human remains have changed greatly from earlier times. The most impactful of those changes has been the rejection of "race science." It is with a real sense of pride and some relief that I can now describe my field of science as being well prepared to engage with and serve all of society. The tools of the humanities, science, and medicine stand ready to provide information about past peoples using informed observation, nondestructive imaging, and tissue samples that are much smaller than those removed during a medical autopsy. Today, osteobiographies benefit from new approaches to the remains of those who died in infancy and childhood, to learn more from them. There are new approaches to skeletons' diverse chemical isotopes and to the paleogenomes of the people and their cohabiting, disease-causing agents.

Osteobiographies can serve an important role in the context of engagement with descendant groups. The repatriation of human remains—that is, the transfer of responsibility for their care—is often appropriate and desirable. The argument needs to be made—clearly and firmly—that scientific study is an important part of wise practice when human remains are discovered, and when archived collections of human remains are to be repatriated. The examples of repatriation activities in both Canada and Africa show that linkages between scholars and descendant groups are crucial if we are to build a respectful, informed future.

Occasional discoveries of human remains will continue. When they occur, scientific exploration should be part of the discovery process. That exploration should be collaborative with descendants of the deceased, if that link can be ascertained. There are many, many human skeletons held in various types of institutions. The past studies of those bones, based on flawed and incomplete biological perspectives, bear little relation to the information and insights that are possible today. For those collections that can be repatriated, fresh scholarship, done in collaboration with descendants, can help to tell ancestors' stories. The repatriation of human remains presents many challenges, as the case studies here demonstrate. It is both similar to and different from the repatriation of culturally important objects. As with objects, human remains can be sources of restitution, empowerment, and new knowledge. There are opportunities for professional organizations, advocacy groups, and governments to find consensus regarding the best approaches, at the global level.

This book is organized around two sections, inspired by my research in Canada and Africa. Each section includes contextual material, examples of osteobiographies, and examples of repatriation. Following a chapter defining key terms and explaining choices about terminology and perspective, the first half of the book focuses on research done in Canada. The two case studies in Chapter 2 provide examples of what transpires when human remains from relatively recent times are discovered. Chapter 3 describes Indigenous cultural, ecological, and regulatory contexts in the Lower Great Lakes of North America, providing the context for the osteobiographies that follow. The case study in Chapter 4 highlights the value of modern paleopathological methods in an exploration of evidence that past peoples sometimes managed to accommodate community members with serious disabilities. Chapter 5 describes how approaches to the bioarchaeology of large collective burials (ossuaries) have advanced, with special emphasis on how collaborative research with descendant communities can provide new evidence about ancestral challenges and achievements. Chapter 6 provides details of the planning and execution of a very large North American repatriation event.

The second half of the book focuses on osteobiographies based in Africa. Chapter 7 introduces African Indigenous and cultural contexts, as well as the colonialism that has influenced past scholarship and collecting. Chapter 8 describes the discovery and interpretation of one skeleton from a group of cemetery sites in northern Kenya, about which relatively little is known. Chapter 9 focuses on southern Africa, delving more deeply into the environment, archaeologically based narrative and history of that region as it pertains to the Khoe-San community. Chapters 10 and 11 provide an osteobiographical overview of the lives of ancient Khoe-San hunter-gatherers, including features of life history and evidence regarding interpersonal violence. Chapter 12 focuses on repatriation in an African context. It includes case studies in which historically known individuals have been repatriated, and other instances in which unnamed persons have been the focus. Finally, Chapter 13 compares and contrasts the repatriation of human remains with the return of objects held by museums. It is proposed that the creation of a more broadly held set of principles is both feasible and desirable.

CHAPTER 1

Osteobiography: A way to see past lives

What is a bioarchaeologist?

I have had a long career as a biological anthropologist—someone interested in humanity's place in the natural world, as seen from a biological perspective. My focus has been the reconstruction of peoples' behaviors—in life—from their skeletal remains (bones and teeth). North American academics refer to this field as "bioarchaeology," Europeans prefer the term "human bioarchaeology." I use the tools of science to expand and deepen humanity's history. I think what I've learned from research in Canada and Africa may be interesting to people.

Some exploration of human origins should be a component of a good, modern education, yet biological anthropologists are so rare that many universities do not have even one on staff. We are rare birds. People ask me how I landed on this branch of the academic tree. My childhood was not like that of the paleontologist who was always fascinated by dinosaurs, or the engineer who was constantly building things. I lived on a prairie farm, one of four children. As I explored the gullies and creek beds, I liked to imagine I was a hunter-gatherer, living off the land, eating a diet of wild plums, cattails and arrowroot. My first nine years of education took place in a one-classroom school with about ten other children, a small wood frame building surrounded by corn fields. Next came a high school with 140 students, set in a farm town of about 800 people. Even in that limited social world, I could see that people didn't always understand one another's points of view. I was bothered by their prejudicial notions about goings-on beyond their own experience. I resolved to become a teacher who could explain people's differences to one another, thereby achieving world peace.

My undergraduate degree was in world religions, since I knew that faith-based ideologies were often flash points. I knew that more training was necessary if I was going to teach about the topic, but my options were limited. A conveniently located university offered a graduate degree in anthropology, which I reasoned might be vaguely similar. Once admitted, I

Osteobiographies
https://doi.org/10.1016/B978-0-12-823880-6.00002-1

was required to take an undergraduate course in physical (as it was known then) anthropology. I was enchanted. The course introduced me to the domain of scholars who seek to understand how and why people and their primate relatives vary in their looks, their genes and their behaviors, through exploration of their origins and natural history. Their work is based on a paradigm: a distinct set of concepts that are scientifically testable. Evolution and its mechanisms, like mutation and natural selection, provide the foundation, but social behavior is a part of every story, too. From that point, I sought training to become that teacher who helps people understand diversity, hoping—even as I matured—to contribute to world peace.

I mentioned that there are relatively few biological anthropology professors in the world. In North American schools, they are usually listed as social scientists. Globally, they can be associated with archaeology, ancient studies, medical schools, or with the natural sciences. This reflects the two-headed creature that inspires their work: Science and History. The methods of science provide information that leads to generalizations about how things work. The methods of history—considered here in its broadest sense—help us understand specific occurrences in the human realm. My focus is on how ancestral populations survived for the immense time span from the origin of modern humans (about 200,000 years ago) to the present day. An example: I may wish to understand how a hunter-gatherer community who lived 2000 years ago cared for a seriously disabled child (see Chapter 4). That is a specific occurrence. The tools of science—like radiocarbon dating, stable isotope analysis, anatomy, palaeopathology and palaeogenomics—provide key information, but those data need to be interpreted within a framework built on assumptions about peoples' social and technological capacities.

What is an osteobiography?

The meaning of "osteobiography" is straightforward. It is the story, or identity, of a person as told through the person's bones (and teeth), supplemented with information about the archeological context, biological, chemical and other analytical information, as available. And, since humans spend their lives in communities, it includes the collective exploration of skeletons that make up a sample of that community. The idea dates back to the 1970s. From that era, a rather simple figure with lots of arrows shows how a researcher could go from questions about the individual to answers about the population (Saul, 1976; Saul & Saul, 1989). Historically speaking, this idea represents a major revision of orientation, replacing the "race science" mandate that dominated the field until just after World War II (Wolpoff & Caspari, 1997). The new idea asserted that the primary goal of studying human skeletons was not their classification into racial groups based on how they looked. Up to this time, because of the preoccupation with racial classification, researchers had often salvaged and catalogued only the well-preserved adult crania from excavations.

For part of my thesis research (Pfeiffer, 1977), I studied human remains from a cemetery that had been used by hunter-gatherers in the Finger Lakes region of New York State about 2500 years ago. The archeological site had been excavated in 1939–40 (Ritchie, 1944). The museum gave me access to all the human remains that the archeologist had brought back from the field. Of the 159 burials excavated, the museum had 96 units. Four were cremations. Of the other 92 units, each of 42 included only an adult cranium and mandible, sometimes accompanied by a long bone. Only eight units were of relatively complete skeletons. Twelve of the units represented children who had died at ages between birth and about 12 years of age. A decision had been made to leave many of the excavated bones behind, prioritizing the retrieval of well-preserved adult crania. Not surprisingly, the initial research done on the collection had focused on how many "sub-racial types" were represented. It had concluded that people with long, high narrow skulls had moved in, and had dominated and replaced the prior occupants who had rounder, broader, lower skulls (Ritchie, 1945). My thesis work showed that the early analysis was incomplete and biased. Statistically, all of the adult skulls grouped together nicely. Ah, those skull-centric days! Those human remains were repatriated and reburied in 2002. Although there were many questions I could not begin to answer from my study, at least the museum was able to archive my raw data and photos for scholars of the future.

The revolt against the idea that ideal races existed and just needed to be found was truly a revolution, re-shaping the scientific paradigm. Researchers who were interested in human evolution encouraged each other to reject the race science mindset and instead embrace a "new physical anthropology." Researchers would interpret population variability in terms of adaptations (Krogman, 1951; Washburn, 1951). Researchers with a special interest in human skeletons from archeological contexts began studying all the bones as a matter of course—not just the nicest looking adult crania. They started asking questions about the lives of children, as well as adults.

This new orientation led to the idea of constructing osteobiographies, as illustrated here (Fig. 1.1). Something of a conceptual black box, on the left side are a number of basic questions rather similar to queries you might

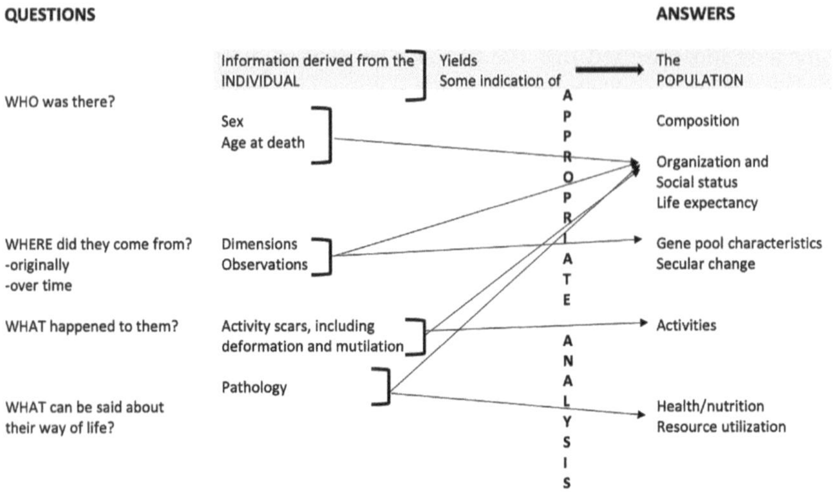

Fig. 1.1 The "osteobiographical" approach. (Modified from Saul, F. (1976). Osteobiography: Life history recorded in bone. In E. Giles & J. S. Friedlander (Eds.), The measures of man: Methodologies in biological anthropology (pp. 372–382). Cambridge, MA: Peabody Museum Press, p. 374.)

make of a new acquaintance who you'd like to know better. Who—in a demographic sense—are you meeting? What is their ancestry? Where have they spent their life and what can be said about the duration and quality of that life? On the right side are features of the "population"—by which is meant the person's community. In the center, positioned like a backbone, is APPROPRIATE ANALYSIS. While quite a few of the specific approaches have changed since the 1970s, the aspiration remains the same, as does the emphasis on doing it right.

This diagram illustrates not only the aspirations, but also the assumptions of those times. The core message is that the study of individual skeletons can tell us about more than just that specific skeleton. The whole, collective past of one person's group can be seen through the sum of the person's skeletal parts. The assumptions and approaches identified in the diagram have changed substantially during the past fifty years. It was generated in the era when the curation of adult crania was prioritized because skulls could provide the best reflection of ancestry through their size, shape and unique developmental patterns (the latter are known collectively as discrete traits). Notice that there is no mention of genomics, or even of DNA. Instead, calipers and checklists were used to quantify the sizes and shapes of bones and teeth (in the diagram as "dimensions, observations"). These were used as proxies for genetic relationships. Those rather imprecise tools can still convey information about ancestry today, but they have been expanded by better measurement tools and supplemented in many instances by palaeogenomics.

The preoccupation with adult crania meant that the information to be gained from the rest of the bones got much less attention. Adult postcranial skeletons and the skeletons of infants and children were of secondary or no interest. While students were normally taught about some tools to estimate approximate infant and juvenile age at death from dental and skeletal development, mastery of that information was quite secondary to the mastery of tools needed to measure, compare and classify mature crania of the world. However, as researchers adopted the osteobiography paradigm it became clear that patterns of child growth, and exploration of the causes behind the deaths of children can lead to important discoveries about the nature and quality of past lives. The study of infants and children has also become very important in the context of forensic anthropology—the study of human remains to resolve legal matters.

Just as important as our improved technology and commitment to studying ALL human remains found in archeological contexts is our shift to a somewhat different set of questions. In the diagram, notice that most arrows connect the individual to the "population" composition, organization and life expectancy. The thinking was that a collection of skeletons, like those found in a cemetery or crypt, was an unbiased cross-section of the community who placed them there. The composition of the collection would tell you about the community's lived experience. For example, if there was a high proportion of infants relative to adults, there were two possible explanations: either there was a high infant mortality rate, or there was a modest mortality rate accompanied by a very high birth rate. If the most elderly skeletons appeared to be no more than 50 years old, then that was the age to which people could live in those days. Researchers were generating palaeo-demographic life tables like those used by demographers in order to understand past communities, but anomalies were accumulating. There were unexplained differences in age-at-death profiles, and a curious shortage of people who had lived long enough to be grandparents.

The osteological paradox

And so, the approach portrayed in Fig. 1.1 came to be questioned from various perspectives. It was shown that the shortage of old people could be explained by age-at-death estimation methods that were biased toward overly youthful assessments. Fine, we could improve our tool kit. The anomalies in other palaeo-demographic features actually represented not methodological, but conceptual problems. By the 1990's, an adjusted perspective had developed, in which researchers were urged to work within constraints that are characterized as the "osteological paradox." The citation classic on this topic has as part of its title, "Problems of Inferring Prehistoric Health from Skeletal Samples." There are three components to the paradox: "demographic nonstationarity, selective mortality, and unmeasured individual-level heterogeneity in the risks of disease and death" (Wood, Milner, Harpending, & Weiss, 1992:343). Stated briefly, we cannot assume that people never changed locations between birth and death; we only get the privilege of seeing the subset who died at each age (not the cohort she/he came from); and we cannot assume that everyone has the same likelihood of dying from a particular disease or natural hazard. Another way of putting this last point is that humans have different levels of frailty. Researchers came to realize that the skeleton with the most extensive osteoarthritis, accompanied by healed fractures and dental caries, might have been the most robust in life, and the skeleton with no "scars of life," might have been the frailest—dying rather than recovering from a stressor that others in the group had survived.

These realizations have shifted the field. The questions asked of individual skeletons are still based on the classic 5Ws: who-what-when-where-why. The answers that can be applied to the community are now oriented more strongly toward genomic composition, geographic mobility, habitual activities, health and disease, staple foods, and various cultural practices that leave traces on the bones and teeth.

Osteobiographies in sociopolitical context

I intend to tell stories that illustrate the value of learning about how people managed, and often thrived, through those many generations prior to modern times. At the same time, I plan to convey the way human skeletons have come to play a role in the re-framing of the past. As dominant forces in modern society reflect on how their actions and assumptions affect others whose voices have not been heard, skeletons—in a way similar to statues, flags, artifacts and other items—can take on contested symbolic importance. The idea that human skeletons from the distant past should not be displayed and should not be held in archives was a firmly marginalized argument until the middle of the 20th century. A gradual shift in perspective was benchmarked by "the Vermillion Accord on Human Remains." This statement, issued by the 1989 World Archeological Congress at a meeting held in Vermillion, South Dakota, recognized the need to balance respect for the wishes of contemporary communities and the scientific research value of human remains. The Accord concludes: "The express recognition that the concerns of various ethnic groups, as well as those of science are legitimate and to be respected, will permit acceptable agreements to be reached and honored" (Pardoe, 2013).

While the Vermillion Accord expressed an aspiration that there be negotiation, the subsequent decades saw many legal and legislative initiatives to solve the "nail" of power imbalance with a big, rather insensitive "hammer." One of the most impactful is a US federal law, Native American Graves Protection and Repatriation Act (NAGPRA), enacted in 1990 (Rose, Green, & Green, 1996). "Repatriation" commonly means return to the country of origin, as when fallen soldiers are brought home for burial. In the case of NAGPRA, the nation is an indigenous community. Under NAGPRA, any institution that receives federal funding is required to seek out Native American lineal descendants and transfer to them all human remains, associated funerary and sacred objects. Thanks to the dynamics of modern life, these materials can only rarely be returned to the physical locale from which they were taken. The idea of transferring something to people who didn't previously know it existed, who may or may not feel affinity toward the object, is conceptually challenging.

Meanwhile, the Australian government developed a Policy on Indigenous Repatriation (IRP) that seeks to return skeletons from Australian institutions to contemporary communities, and an International Repatriation Program that seeks the return to Australia of bones and sacred objects from oversees institutions. Other countries have launched initiatives as well. All this has been disconcerting to the people responsible for the collections. Trained or identified as curators, they have in recent times been designated to document, coordinate and physically accomplish new mandates. Institutional budgets have added new line items, although not necessarily new funds. Sometimes the receiving community has been receptive and prepared, sometimes not.

The rationale for action is in some ways very clear. There are examples from many parts of the globe in which the recently dead—named individuals—were "procured" from graves and curated by 19th century institutions in the name of race science (Gould, 1981; Legassick & Rassool, 2000). Later chapters in this book document some of these cases. These actions are indefensible when seen through a modern lens. Twentieth century commentary often extended condemnation to include more recent, purposeful excavation of previously undocumented cemeteries. Early in my studies, I was a crew member on one of these excavations. We removed the bones of over 500 people from their collective resting place. It was exciting, educational and unsettling work. I am happy to say that 43 years later the bones were reinterred in pretty much the same location, but more on this later (see Chapter 6). Since the excavators in such cases came from a privileged part of society and the skeletons were the ancestors of the marginalized, these actions were insensitive at the least. It can be argued that they were unethical, although they were not perceived as such at the time. The fates of the many, many skeletons that have been disturbed by accident—building projects, road widenings, drainage repair—have been swept up into the general mood of injury. It is difficult to decide where to stop, in framing this broader clash of perspectives. It shares space with arguments for return of "cultural properties/objects of cultural patrimony" (see Chapter 13) and it links to discussions along a similar line regarding genetic research. The core issue is one of social privilege—perceived and real.

It is no surprise that counterarguments about the scientific value of research done on human remains and artifacts were not immediately effective. After all, there is no doubt that the skull collectors of the 19th century were pursuing an errand that was both unethical and ill-conceived. The idea of prototypical "pure" human races is now well and truly dead. Today, those collections document the follies of the past. On the other hand, there are limited avenues for learning about the lives of the millions of people who represented our species for thousands of years. It has been argued that reburying their skeletons is an act of willfully denying potential new knowledge (Weiss & Springer, 2020). Arguments against the return of cultural objects follow a similar line of thought. The directors of twenty of the world's most prominent art museums have crafted a "Declaration on the Importance and Value of Universal Museums." Their stance on the repatriation of objects, while nuanced, comes down on the "no" side: "Although each case has to be judged individually, we should acknowledge that museums serve not just the citizens of one nation but the people of every nation."

Thirty years since the passage of NAGPRA, there has been considerable progress toward collaboration between researchers and representatives of the ancestors, not just in the USA, but around the world. Although the American legislation has become well known, most countries have not mandated broad handovers of museum collections through legislation. Nevertheless, the sentiment in support of conveying responsibility for skeletal collections to living persons who are in some way related to them has taken hold. There has been progress in many jurisdictions toward better zoning laws so that archeological sites (including cemeteries) are protected, and toward heritage legislation that restricts excavations, encourages consultation and collaboration. The professional associations representing researchers like me have adopted strongly worded ethics statements. Gradually, alliances are being formed and collaborations are being pursued. When I began teaching, skeletons were given cheeky nicknames and human bones were used for practical jokes. Such cavalier devaluation of past human lives is now much less common.

In this book, I will offer examples from my experience of the ways we can learn from the bones and teeth of past people. All stories will include information on how those remains came to be available for study, what we were able to learn about them, and what happened to the remains after we studied them. In my view, there is always value in pursuit of documentation and research when human remains are encountered. Just as forensic scientists have a serious responsibility to "speak for the dead," so bioarchaeologists have a responsibility to learn about our shared past through the stories of those who came before us. As society moves toward a world in which heritage issues are handled in a balanced, respectful way, people need information about what "research value" means. This book is a small step toward providing that information.

The tools of the trade

The study of human skeletons relies on very different lines of evidence. Some are classically objective. They are quantitative and replicable. Others are classically subjective. They are normative, subject to interpretation. The mingling of methods is both exciting and challenging. All kinds of interesting, testable hypotheses can be created. Nevertheless, the evidence of the past is often shrouded in uncertainty.

The science side

When Aleš Hrdlička established the *American Journal of Physical Anthropology* in 1918, peoples' understanding of the natural world was based on how things looked. Hrdlička, a childhood immigrant from Europe to the U.S., was the first curator of physical anthropology at the U.S. National Museum, now known as the Smithsonian Institution National Museum of Natural History. He was responsible for collecting and cataloging the skeletons of many Native Americans, as he explored theories about their origins. Not much advanced from the tools available to Darwin fifty years prior, scholars of the early twentieth century who wished to understand how humans fit into natural history studied the similarities and differences in the physical characteristics of humans and their likely close primate relatives: apes, monkeys and prosimians. The classification of humans into subgroups (races) was built on measurements and categorizations for traits like nose shape, lip thickness, and—of course—skin color.

Physical anthropology was "the study of racial anatomy, physiology and pathology," according to Hrdlička. Like him, many of the early practitioners were trained in medicine. Indeed, knowledge if musculo-skeletal anatomy is a fundamental and enduring tool for researchers. They need not only to recognize fragmentary remains, but also to know what variations represent normal difference. For example, the adult human breast plate, or sternum, is formed from several smaller centers of bone formation. Rarely (but perfectly normally), they don't join completely. This leaves a smooth-edged hole along the midline. Every now and then, a naïve observer will decide that such a perforated sternum represents a healed gunshot or stab wound. Along the same line, mistakes can occur when a discoverer can't distinguish the deciduous dentition from the permanent, thereby misclassifying the fragmentary remains of a child as adult, and vice versa (Box 1.1). Knowledge of skeletal and dental anatomy, not just in the adult but also throughout the lifecourse, is a foundational tool.

BOX 1.1

A human skeleton is made up of several kinds of tissue, each of which reflects a lifetime of development, maintenance, and deterioration—the latter being a privilege, if we live long enough! Each stage, but especially development, is programmed and controlled by the person's genome, or complete set of DNA. As fetal cells differentiate, the building of bones and teeth begins. While quiz games normally peg the number of bones in the skeleton at 206, almost every one of those bones is made of multiple, independent ossification (bone formation) centers that fuse at some point between early fetal development and young adulthood. By the 10th week of gestation there are over 800 points of bone formation; by the time of birth there are more than 450. At birth, collagen templates and subsequent mineralized bone tissue are present throughout the body. Some regions—like the wrist bones—are present only in their collagen forms. There are seven primary bones in the human wrist, but an x-ray will show fewer than this until mid-childhood, when all have finally at least begun to ossify.

The dentition, similarly, is a hotbed of activity early on. Arrangements must be made for 44 tooth buds, representing the 20 deciduous and 24 permanent

BOX 1.1—cont'd

teeth. Each tooth is made of enamel, dentin, and cementum. The latter two tissues are made of collagen and mineral, as is bone tissue, but they are denser. In life, bone tissue is about 20% water, dentine and cementum are 4%–5% water. Enamel is a unique tissue. It is not enervated. Once formed, it has no living cells, it incorporates almost no water, and it has a crystalline structure. It is like rock, despite its biological origin.

The chemical composition of each type of hard tissue captures aspects of a person's life during growth. After growth is completed, bone continues to remodel, capturing information about the latter part of a person's life. Hard tissue chemistry is not the only personal identifier. More obvious, and more central to traditional studies of human remains, are the distinguishing surface characteristics, plus the sizes and shapes of individual bones. The guessing-game number of 206 is an example of false precision, as the numbers of ribs may vary, the hands and feet may have tiny bones within muscle sheaths, the skull may have small ossicles along suture lines, and so forth. There are variations in the placement of holes in bones (called foramina), prominence of muscle attachments, and the mass of individual bones. The variations are remarkable, considering the firm control within which the bones develop.

The tempo of a skeleton's development is dictated by evolutionary factors that have shaped the genome (Bogin, 1999; Sinclair & Dangerfield, 1998). Illustrated in a very basic way by R.E. Scammon in 1930 (Scammon, 1930), Fig. B1.1 emphasizes the different drumbeats that are synchronized between birth and maturity. The lymphoid curve expands to twice its adult capacity, to protect the growing, developing child from pathogens. The neural curve tracks the very rapid expansion of the brain and central nervous system. This is reflected in the growth of the cranial vault and the neural canals of the vertebrae. By age 7 years, the brain is almost fully adult in size, and the bones protecting the skull are large as well, although not as thick as they ultimately will be. The general curve, sometimes called the sigmoid curve, tracks the growth of most bodily systems, including muscle mass, stature, and long bone lengths. The child's size increases rapidly, then more slowly in mid-childhood, then rapidly again during puberty. It is only at the end of puberty that bone growth centers can fuse, since those final fusions act to end linear growth. The genital curve is quiescent until puberty. Hormonal changes then stimulate extensive changes, reflected in the skeleton by changes to female pelvic shape, increased shoulder breadth and a more pronounced chin in males. It is very difficult to determine the sex of a skeleton until after the genital growth spurt has commenced. Skeletal features that happen during growth are more predictable than features that happen later. Skeletal features like bone mass, pelvic responses to childbirth and osteoarthritis reflect the adult years in which genetic control gives way to the effects of daily behaviors, diseases, environmental hazards, and aging.

Continued

BOX 1.1—cont'd

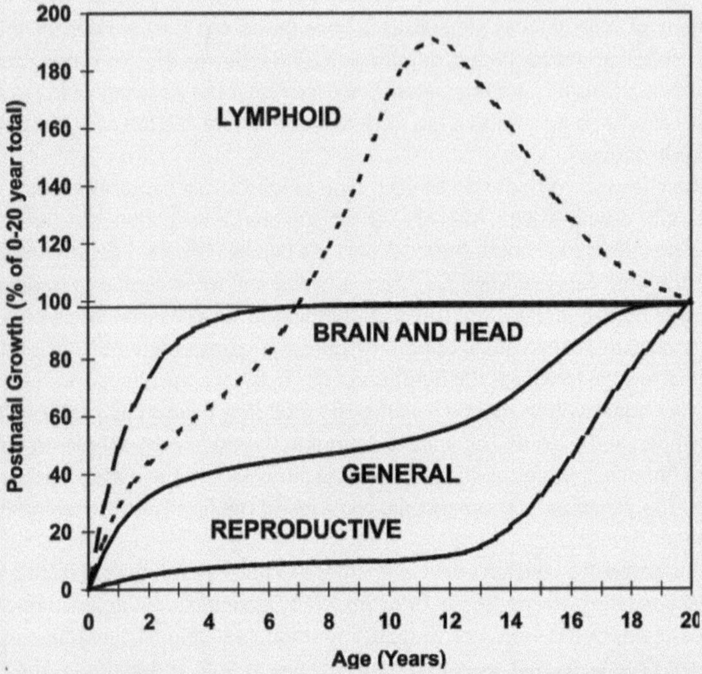

Fig. B1.1 Scammon growth curves showing different parts and tissues of the body. All curves are of size attained are plotted as a percentage of total gain from birth to 20 years. Size at 20 years is 100% on the vertical scale. (Originally from Scammon, R. E. (1930). The measurement of the body in childhood. In Harris, J. A., Jackson, C. M., Paterson, D. G., & Scammon, R. E. (Eds.), *The measurement of man* (pp. 171–215). Minneapolis, MN: University of Minnesota Press; reproduced from Lichten, W. (2004). On the law of intelligence. *Developmental Review, 24*(3), 252–288.).

In North America, the field gradually became recognized as distinct from medicine, and the Ph.D. identified as the ultimate credential. As the scholarly group dedicated to the study of human evolution, physical anthropologists were placed in departments of anthropology at many universities. By the middle of the twentieth century, their roles had changed. Race science was finally being rejected and the tool kit for exploring diversity between and among humans and their relatives had expanded greatly. No longer were all studies based on how things looked. Gradually, the label "physical" anthropology was supplanted by "biological" anthropology. As of 2020, the world's largest professional body for the field, the American Association of Physical Anthropologists, is named the American Association of Biological Anthropologists.

This bit of history helps to explain the transformation of the tool kit used by practitioners of the field. It all began with calipers. They were designed in many sizes and shapes, to measure lengths, widths and angles of skulls, long bones and teeth. The quest in those days to find an accurate way to measure the size of a person's brain based on their skull size and shape led to some unique hardware. If you want to see some remarkable inventions, just type "craniometer" or "craniophore" into an online search engine. Calipers of more familiar dimensions are still useful—both sliding and spreading are worth packing in the tool bag—but they have been joined by tools that cannot be slipped into a backpack. Many of today's tools require the bone or tooth to come to the lab for three-dimensional imaging or for chemical analyses. As noted by Harper in *The Fate of Rome*, "the isotope chemistry of bones and teeth can tell stories about diet and migration, biological biographies of the silent majority"(Harper, 2017:13).

The fact that most elements of the periodic table exist in both common and isotope forms has been a boon to researchers who study skeletons from the past. An isotope is a form of an element that contains different numbers of neutrons in its nucleus. It differs in mass but not in chemical properties. Some isotope forms are stable, others are unstable. The unstable forms "decay" at a predictable rate. Carbon, that very common element in all living things, normally has six neutrons and six protons, but the number of neutrons can vary from three to nine. The superscript before the element symbol is the number of neutrons plus protons, which for carbon ranges from 9 to 16. These are referred to as isotopes of carbon. It is the unstable ^{14}C isotopic form of carbon that is frequently used for determining antiquity in archeological settings. Deterioration of other elements' unstable isotopes that become trapped in solidifying lava or in deeply buried rock strata can be used to determine the antiquity of much older palaeontological events (Box 1.2).

Measurement of antiquity is a good starting point, but there are additional tools for researchers, thanks to the stable isotopes of various elements. Because they differ in mass, isotopic forms often take slightly different pathways in natural processes, like metabolism. The story of ^{13}C will illustrate their potential. From its discovery, ^{14}C dating was very popular with researchers. Lots of samples were being analyzed. In order to determine the

BOX 1.2 Radiocarbon dating

Life on Earth is carbon-based. It makes sense that carbon is a very closely scrutinized element: It is us. The carbon atom normally has twelve neutrons, but this number can vary. Forms of an element with the same number of protons but different numbers of neutrons are called isotopes. A stable isotopic form, ^{13}C, makes up about 1% of global carbon. It is important to dietary studies. The unstable or radiogenic form, ^{14}C, is valuable for determining the antiquity of formerly living things. Formed from ^{12}C exposed to cosmogenic radiation, a small amount of ^{14}C is incorporated into all living things through processes like respiration, tissue formation and maintenance. When an organism dies, those processes cease. At that point, the carbon is trapped, the ratio of radiogenic to stable carbon is fixed, but the ^{14}C continues radioactive decay while the ^{12}C and ^{13}C do not. The half-life of ^{14}C (the period of time after which half of a given amount will have decayed) is about 5730 years. This means that after about 50,000 years, there is none left. The range of time within which ^{14}C dating can be applied is thus from the recent era to about 50,000.

Since the 1940s, researchers have improved many aspects of radiocarbon dating. Early discoveries included measurement of the impact on dating from above-ground nuclear bomb tests in the 1950s and 60s, and from the global burning of fossil fuels. Other factors that create deviations in the calibration curve are atmospheric and planetary activity, as well as the geographic source of the carbon taken up by the organism being measured. The latter is known as the carbon exchange reservoir. For example, oceans suck up carbon and may hold it for long periods, so marine organisms may yield ages that are older than is true. All this is considered within the calibration curve that is applied to the measured ratio of ^{14}C to ^{12}C. Adjustments to the calibration curve are frequent. Evidence from tree rings, lake and ocean sediments, corals and stalagmites are used to fine tune its accuracy and precision. It was last updated in 2020 (Jones, 2020).

Progress is also regularly made on the amount of material and the type of material that can be tested. The early observations were made with Geiger counters, but since the 1970s, measurements come from accelerator mass spectrometry (AMS). Modern laboratories can generate dates from very small samples of traditional sources like charcoal and bone, and a remarkable array of organics that include soil, seashells, ivory, textiles, individual seeds and grains, straw from within mud bricks.

The standard reportage of AMS results includes the uncalibrated date in years before present (BP)—1950 is the zero date—and an estimate of error to 1σ. The calibration, taking into account the sample's organic type and geographic origin, adjusts both the date and the error term. Dates are normally reported as cal BP, cal BC or cal AD. The calibration program and version are also reported, as well as the laboratory designation. Radiocarbon dating is very important to many fields, including archaeology. Often archaeology students are surprised to learn that part of the required curriculum is taught by a physicist. While this brief synopsis cannot do it justice, it establishes the linkage between ^{14}C dating and fundamental Earth processes. Radiocarbon dates are facts, not opinions.

Jones, N. (2020). Carbon dating, the archaeological workhorse, is getting a major reboot. *Nature*. doi:https://doi.org/10.1038/d41586-020-01499-y.

proportion of ^{14}C in tissues from the past, researchers must measure its abundance relative to the stable forms of carbon: ^{12}C and ^{13}C. It was noted that the ratio of those two stable isotopes was variable. In North America prior to European arrivals, hunter-gatherers ate foods derived from naturally occurring trees, bushes and animals until a plant domesticated first in Central America was broadly adopted: maize. In Mexico, maize had been dramatically modified from a form of grass. Dating labs noticed that the bones of people who had eaten a lot of maize had a higher ratio of $^{13}C/^{12}C$ than people who hadn't. This is because, generally, grasses have a metabolic pathway different from trees and bushes. Photosynthesis—the series of chemical reactions through which most plants transform sunlight, carbon dioxide and water into the nutrients they need—can take place through various metabolic pathways. Grasses prioritize rapid growth and are less discriminating against the heavier ^{13}C isotope.

Dinner's plant-based course is not the only menu item that leaves a clue. A difference can be discerned if the person of interest mostly ate the meat of animals that ate grass (grazers), or animals that ate trees and bushes (browsers). While the topic of dietary reconstruction through stable isotope patterns got its start in North American archaeology, it now is a global field of research. There are dietary differences that can be discerned through nitrogen and sulfur isotopes, and there are environmental differences that can be discerned through oxygen and strontium isotopes. I will refer to these topics as they arise in subsequent chapters, but I will barely scratch the surface. The impact of stable isotope geochemistry on the study of the human past has been huge.

Palaeogenomics—the study of the past through the examination of preserved genetic material from the remains of ancient organisms—also belongs under the heading of "huge." After an organism dies, all manner of physical and chemical changes occur. Some tissues decompose quickly, others slowly. The pace and pattern of decomposition are strongly influenced by the temperature and humidity of the surroundings. Preservation will be worst in warm, wet settings. Places that are very dry, or dry and consistently cold are best. Preservation is also affected by the way cells are structured. Some of the first successful extractions of cells from dead humans were of *Mycobacterium tuberculosis*, because of its very tough protective membrane. This disease vector is, of course, extrinsic to the deceased person—even if tuberculosis (TB) was the cause of death. As laboratory methods have quickly improved, extraction of intrinsic cells has become possible. Mitochondrial DNA was an early win because it is very common within all human bodies. Nuclear DNA was more challenging, but it now can be extracted from some—but far from all—human skeletons. Our perspective on past genomes is currently most complete for skeletons from cold, dry places like Ice Age Europe and northern mountainous regions. It is a wonderful tool, with great potential. Perhaps extraction of ancient DNA will become as common as radiocarbon dating, but for now, it is an expensive initiative that often fails—especially when skeletons come from moist, temperate or tropical soils.

The humanities side

Scholars who are devoted to understanding Biblical or Koranic times focus on documents and archaeology. The same is true for scholars who focus on ancient civilizations of China, Greece, Rome, the Viking Age and so forth. However, most of the world's past peoples did not leave behind epic poems, hieroglyphics, Dead Sea Scrolls, or any other documents. Even the identities of people who lived in recent times may be completely untraceable if their lives were not documented or archives were destroyed (Mant & Holland, 2019). If we wish to reconstruct the lives of those who lived in the distant past, there are four potential sources: historical documents made by literate visitors, ethnographies of modern descendants, descendant communities' stories of their ancestors, and archaeology. Each of these story lines has its strengths and weaknesses. It is rare indeed for all four lines of evidence to be available in any one case. In my line of work, the only constant is archaeology. The quality of the information that comes from excavations can be poor or excellent, as the case studies in later chapters will demonstrate. The same is true for the other lines of evidence.

With regard to history, I can do no better than to quote Jake Silverstein, editor in chief of the New York Times Magazine:

> *"Historical understanding is not fixed; it is constantly being adjusted by new scholarship and new voices. Within the world of academic history, differing views exist, if not over what precisely happened, then about why it happened, who made it happen, how to interpret the motivations of historical actors and what it all means"*
>
> ***(Silverstein, 2019:8).***

So true. Colleagues who are trained in historiography—the study of historical writing—are important to the task of reconstructing past lives.

Ethnography is the scientific description of the customs of individual peoples and cultures. The same term is applied to the documents produced from that research. An ethnographer is at least to some extent an outsider to the peoples she seeks to understand. There is a vigorous methodology, and ethnographies are subject to critique in much the same way that histories are. In my work, some very valuable documents blur the line between history and ethnography, being first-person observations by explorers and missionaries to parts of Canada and Africa. On the one hand, they sought to describe "customs," but they did so with the eyes of outsiders who had specific priorities and biases—both explicit and implicit. They provide the most information about religious beliefs and about individuals who they perceived to be the "headmen"—indeed, men in general. Food gets some attention, but other important features of life—like childbirth and child rearing—are generally ignored. Although they may be biased, we are lucky to have them. They must be interpreted with care. As one goes further back in time, the value of these sources diminishes. For the deep past, archaeology is all there is.

Terms, labels and designations

Stories about past communities rely on constructed terms to convey actions and identities. This gets us into the thorny landscape of labels. Stories of past groups of people are full of labels that are less than ideal. Sometimes the labels began as derogatory terms used by neighboring groups. Sometimes they are based in geographic confusion, like the term "Indian" for American indigenous peoples. In Canada, the Constitution Act defines "aboriginal peoples of Canada" to include "Indian, Inuit and Métis peoples," yet terms like First Nation and Indigenous group are generally preferred by contemporary "Indian" groups in Canada. The choice of the best name for their ancestors should be made by the descendants, but this is not as simple as it sounds. Descendants can be broadly dispersed—geographically and ideologically; they can speak diverse languages and can have diverging views about what terms are best. Views can shift, as they have with regard to the southern Africans known as Bushmen, a.k.a. San, Khoisan, Khoe-San, and other variants. In this book, I try to use terms that, to my knowledge, are currently preferred by descendants. In that spirit, I will use the overarching term "Indigenous," capitalized (Weeber, 2020), when speaking of past people to whom I cannot give more specific names.

There is a long scientific tradition of being very reticent about proposing links between people of the distant past and the present. The intent in this, I think, is to avoid the suggestion that cultures and behaviors don't change. It is sometimes argued that if an archeologically defined tradition is linked to a living group, this somehow diminishes the extent to which its features can be generalized globally, for understanding the human past, writ large. In that spirit, past communities have been labeled by their archeologically defined culture categories rather than labels linking them to contemporary descendants. This leads to labels like Middle Woodland people or Later Stone Age communities. While these labels are factually accurate, they enshrine past peoples and events in a separate world, disconnected from people alive today. Archeologists and geneticists have made considerable progress in making the connections between past cultures and living descendants. In this book, I will do my best to identify who is ancestral to whom. I will write about ancestral Huron-Wendat people in lieu of calling them Late Middle Woodland people. The ancestral Khoe-San people of Southern Africa are associated with a culture complex known as Later Stone Age. By using the term ancestral Khoe-San, I will give priority to their lineage—which continues—over their past culture—which doesn't. Archeological evidence of ancestors does not provide a static document, but rather shows us a complex mix of elements, some more stable through time than others. No contemporary or historic group is unchanged from its many ancestral iterations.

The baggage of history is also reflected in place names. Heritage regulations frequently impose code names to archeological sites but the sites almost always have more commonly used names that are assigned by discoverers or excavators, as a kind of privilege. Those names reflect contemporary land use—like a neighborhood or creek name—or the name of the landowner. Here are some examples from Ontario, Canada. One of the first sites I helped to excavate was named after the farmer on whose land it lay. His last name being Hind, it is known, indeed, as "the Hind Site." Very rarely, an archeological discovery can be linked to a place of significance in the record of the Indigenous people who created it. This is the case for the Ossossané site, created by Wendat people in 1636 CE, named in the account of a European observer. A recent trend is to re-name sites after important Indigenous people. In 2011, a large Indigenous village site had been named the Mantle site, after the landowner (Birch & Williamson, 2013). It was re-named the Jean-Baptiste Lainé site. The new name commemorates a decorated WWII Huron-Wendat veteran. In 2013, a southwestern Ontario burial site that had been known as the Kleinburg Ossuary since its discovery and excavation in 1970 was re-named Thonnakona. The new name commemorates a very important leader who met and traveled with the explorer Jacques Cartier, from 1534 CE. It is important to remember that most sites and associated collections with Euro-centric names actually represent the lives and actions of people whose languages and customs differed substantially from the world evoked by site names.

References

Birch, J., & Williamson, R. F. (2013). *The Mantle site: An archaeological history of an ancestral Wendat community*. Lanham, Maryland: AltaMira Press.

Bogin, B. (1999). *Patterns of human growth* (2nd ed.). Cambridge: Cambridge University Press.

Gould, S. J. (1981). *The mismeasure of man*. New York: W.W. Norton.

Harper, K. (2017). *The fate of Rome: Climate, disease and the end of an empire*. Princeton and Oxford: Princeton University Press.

Krogman, W. M. (1951). The scars of human evolution. *Scientific American, 185*(6), 54–57.

Legassick, M., & Rassool, C. (2000). *Skeletons in the cupboard: South African museums and the trade in human remains 1907–1917*. Cape Town: South African Museum.

Mant, M. L., & Holland, A. J. (Eds.). (2019). *Bioarchaeology of marginalized people*. New York: Elsevier.

Pardoe, C. (2013). Repatriation, reburial and biological research in Australia. In L. Nilsson Stutz, & S. Tarlow (Eds.), *The Oxford handbook of the archaeology of death and burial* (pp. 733–762). Oxford, UK: Oxford University Press.

Pfeiffer, S. (1977). *The skeletal biology of Archaic populations of the Great Lakes region. Vol. 64*. Ottawa: National Museums of Canada.

Ritchie, W. A. (1944). *The pre-Iroquoian occupations of New York state. Vol. 1*. Rochester, NY: Rochester Museum of Arts and Sciences.

Ritchie, W. A. (1945). *An early site in Cayuga County, New York: Type component of the Frontenac focus, Archaic pattern. Vol. 7*. Rochester, NY: Rochester Museum of Arts and Sciences.

Rose, J. C., Green, T. J., & Green, V. D. (1996). NAGPRA is forever: Osteology and the repatriation of skeletons. *Annual Review of Anthropology, 25*, 81–103.

Saul, F. (1976). Osteobiography: Life history recorded in bone. In E. Giles, & J. S. Friedlander (Eds.), *The measures of man: Methodologies in biological anthropology* (pp. 372–382). Cambridge, MA: Peabody Museum Press.

Saul, F. P., & Saul, J. M. (1989). Osteobiography: A Maya example. In M. Y. İşcan, & K. A. R. Kennedy (Eds.), *Reconstruction of life from the skeleton* (pp. 287–302). New York: Alan R. Liss.

Scammon, R. E. (1930). The measurement of the body in childhood. In J. A. Harris, C. M. Jackson, D. G. Paterson, & R. E. Scammon (Eds.), *The measurement of man* (pp. 171–215). Minneapolis, MN: University of Minnesota Press.

Silverstein, J. (2019). The thread. *The New York Times Magazine*. 12.29.19.

Sinclair, D., & Dangerfield, P. (1998). *Human growth after birth* (6th ed.). Oxford: Oxford University Press.

Washburn, S. L. (1951). The new physical anthropology. In *Transactions of the New York Academy of Sciences* (pp. 2–8). New York.

Weeber, C. (2020). Why capitalize "Indigenous"? *Sapiens*.

Weiss, E., & Springer, J. W. (2020). *Repatriation and erasing the past*. Gainesville, Florida: University of Florida Press.

Wolpoff, M., & Caspari, R. (1997). *Race and human evolution*. New York: Simon and Schuster.

Wood, J. W., Milner, G. R., Harpending, H. C., & Weiss, K. M. (1992). The osteological paradox: Problems of inferring prehistoric health from skeletal samples. *Current Anthropology, 33*, 343–370.

CHAPTER 2
Rest in peace

Prospect Hill

Someone had cared a great deal about this young woman. That she had died tragically was obvious (Fig. 2.1). The size and shapes of the tiny fetal bones in her pelvis meant that she would have appeared pregnant but was far from full term. Surely the pregnancy and her youth must have motivated the partial autopsy that had been done (Fig. 2.2). But why did that examination include examination of the brain and exclude the opening of the torso? It had happened only about 150 years ago, in a town of about 1400 people that had seven churches, some schools and a daily newspaper. Why couldn't we give her a name?

This was only one of many frustrations we faced while investigating about one hundred grave shafts on an urban lot in a small Ontario town, back in 1989. An entrepreneur had purchased the lot from the city after the public school there had been closed. Construction of a new assisted living facility was well underway when workers discovered human bones. Attention shifted to the land's use before the school was built. This had been a cemetery for a Wesleyan Methodist congregation, in use from 1824 CE to 1879. It was closed in 1882 when the land was sold. A notice in the newspaper (*Newmarket Era*, 08/04/1882) urged, "Parties who have friends interred in the rear of the plot should have the bodies removed to the new cemetery at an early date." A public school was built on the church plot, with the cemetery area located beneath the playground and driveway. Headstones and other markers were gradually removed or destroyed—although we did find the smooth marble backside of one headstone being well used for the making of taffy and fudge in a local candy shop. Townspeople said others had been used as baseball bases.

What happens when human remains are discovered? There are laws, bylaws and regulations for that. They vary regionally, of course, but they usually direct someone to pay for someone to remove the bones before activities resume. Here, action was dictated by the Ontario Cemeteries Act. The developer was required to hire archeologists to determine the scope of

Osteobiographies
https://doi.org/10.1016/B978-0-12-823880-6.00003-3

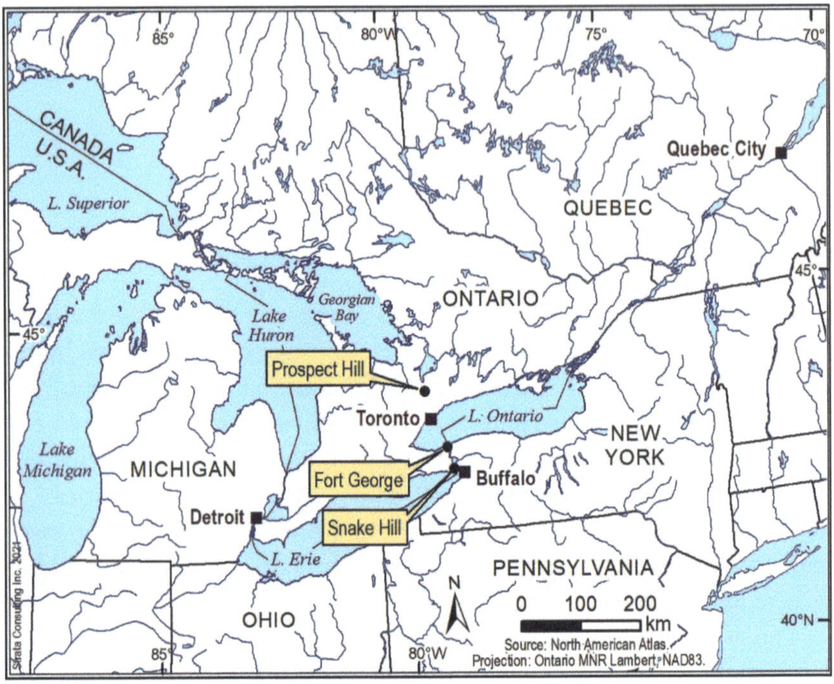

Fig. 2.1 A map of the region discussed in this chapter.

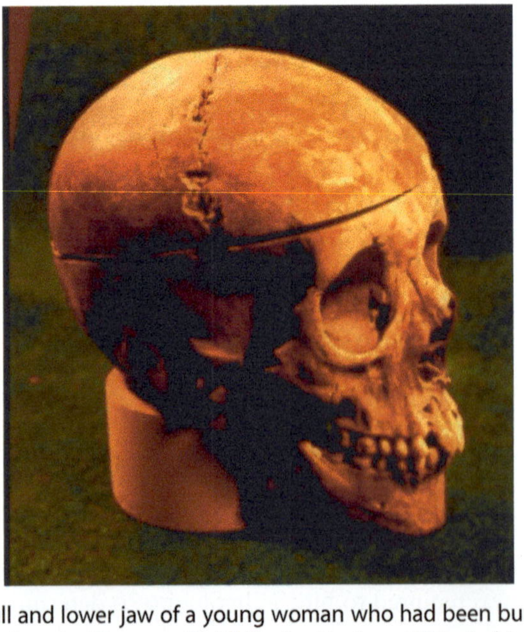

Fig. 2.2 The skull and lower jaw of a young woman who had been buried at Prospect Hill, Newmarket, Ontario. *(Photo credit: J.C. Dudar; Northeast Historical Archaeology, 1989, vol. 18, page 44.)*

the problem and mitigate it. Judging from the grave shafts, fewer than a third of the cemetery's occupants had been removed, back in the day. Seventy-seven skeletons lay in the remains of simple wooden coffins, aligned east-west with heads to the west. Some had been disturbed by installation of school utilities, decades earlier.

A few cold, wet autumnal weeks were scheduled for their removal by a team from Archeological Services Inc., Toronto (Fig. 2.3). We were not allowed to remove any bones from the site, and there was no source of running water to clean off the ever-present mud. I was responsible for planning how to maximize the information we could get from the bones and teeth as they were disinterred. It was a rushed but not quite frenzied time. Survey, photography, line drawings and identification of all artifacts (mainly coffin nails) proceeded. Bones were removed systematically, with special attention given to the teeth because they offer the most information about age at death. By February 1, 1990, the job was done. Construction could resume. With permission of the provincial agency, a bit of rib bone from each skeleton was set aside for later tests. Twenty graves in one corner of the property were not directly threatened by construction, and so were left alone. Meanwhile, authorities responsible for the municipal cemetery declined to provide space for interment of more bodies. Despite vigorous published requests, no family members came forward, no sectarian group expressed willingness to claim these skeletons as their own. It was decided that the corner of the property where some bodies remained would be given cemetery zoning. Each of the 77 skeletons was placed in a numbered black bag, they were nestled together inside a concrete culvert section, and were thusly re-interred. Today, there is indeed a retirement home on the lot, the Chartwell Alexander Muir Retirement Residence. I suspect that the cemetery in the corner of the property is of no consequence to residents. End of story, except for all the interesting things we learned from those 77 skeletons.

Historical evidence

The coffins at Prospect Hill were odd. As the excavation proceeded, we were not finding what we expected to find. The prior year, a nineteenth century family plot had been excavated less than 100 miles away (Saunders & Lazenby, 1991). The fifteen coffins of the Harvie family had been impressively ornate. Some had plaques fastened to the lids with name, birth and death dates. Some had glass viewing windows in their lids, on which woodland scenes had been painted. The bodies lay in cloth-lined boxes, with chaff-filled cloth pillows beneath their heads. Not only were there ornate

Fig. 2.3 Excavation at Prospect Hill, Newmarket, Ontario, 1989. *(With permission of Archeological Services Inc.)*

coffin handles—swing bail and bar types—but also escutcheons (flat metal plates used as guides for thumbscrews), decorative thumbscrews and caplifters. Most coffins were hexagonal in shape, known as the pinch-toe type. The dates of use for that cemetery aligned closely with those of Prospect Hill. The Prospect Hill coffins existed only as iron nails and fragments of pine wood. One coffin was encased in a brick crypt. Only one coffin had handles. Where was everything else?

One possibility was that the simple coffins were those of paupers. There certainly were paupers' cemeteries in nineteenth century North America, but they are normally found near historic hospitals or charity homes. We knew this was a church yard. Could members of this congregation have been of such low economic status that they could not afford fashionable coffins? The answers came from the work of a historian with a special interest in all things Methodist, Marilyn Färdig Whiteley. Most people in the community traced their ancestry to the British Isles, as was common in this part of Canada. Documents from the era are scant; there were not yet published obituaries. Her study of village census data confirmed that these were townspeople with diverse skills. The alphabetical lists run from baker, barber, blacksmith to shoemaker, tailor, wheat buyer, wheelwright. Although not the wealthiest sector of the community, they were an active force in local clubs and organizations. Town records from the period list several physicians and a dentist. The community had access to medical care and education. From long bone lengths, we estimated the stature of each skeleton and noted that the estimated coffin lengths corresponded nicely. This suggests that local carpenters were constructing coffins on demand. A small number were pinch-toe shape, so that fashion trend had not escaped them. Why no fancy do-dads? The answer comes from John Wesley himself, as reported in the *Christian Guardian* (15 May 1872): John Wesley set an example to his fellows by ordering that at his burial there should be no pomp, no equipage, no escutcheon; he would be attended to his grave only by the tears of those who were following him to "Abraham's bosom."

The people buried at Prospect Hill were adherents to Wesleyan Methodist doctrines with respect to burial. That is reflected in their very simple coffins (Whiteley, Pfeiffer, Austin, & Dudar, 1993). We could proceed with other research, assuming that the people we were studying were pursuing trades and living what in those days would have been considered a middleclass life. While we could not document their names, sexes, dates and causes of death, we could perhaps develop some sense of the arc of their lives. A key point of reference for such work is each person's age at death.

Biological evidence

From soon after conception to well into adulthood, a person's teeth are almost inflexible in the tempo of their biological development. There are two sets—deciduous (baby) and permanent—within which different types of teeth each march to their own biological drummer (Box 2.1). Development begins with the tip of the crown and extends to the tip of the root. Enamel and dentine tissues are formed by specialized cells. It is a wonder. If there are serious disruptions to the person's health, the quality of the dental tissues may suffer, but the timing of formation and the size of the tooth are usually not affected. As Prospect Hill skeletons were removed from the mucky soil, J. Christopher Dudar and I used a bucket of water and toothbrushes to clean the jaws enough to assess the developmental stage of each tooth. We also noted which teeth had caries (cavities) and which bony sockets had closed, indicating that teeth had been lost during life. Since the socket gets rather quickly filled in with bone, this change is apparent, and it can be a helpful observation when the skeleton is of someone beyond middle age. The ages of older adults are very difficult to assess, except as broad ranges.

Almost half of the skeletons were of infants and children. Fifteen were infants who had died during the first year of life. Another seven had died before their second birthdays. There are many factors that can cause infant death, most of them so quick to act that bone tissue does not have time to respond in a way we can see. However, two of the infants showed the sort of changes to the skull bones that can be a sign of scurvy (deficiency of vitamin C). Combined with the fact that every child with erupted teeth had at least one cavity, we wondered about the adequacy of their diets. No one—adult or child—had a dental filling. It seems that removal of the offending tooth was the main strategy used by the community dentist. Although there were many children, there was also evidence that long, healthful lives were possible. Some of the adults were old enough to have had joint abnormalities, back problems, low bone mass, and to have been fitted with dentures. They were not buried with their dentures, but prolonged pressure causes characteristic changes to the bone of the jaws (Pfeiffer, Dudar, & Austin, 1989).

Exploring the question of peoples' preferred diet further, we explored chemical features of those small rib bone samples we had retained. In her lab, my colleague M.A. Katzenberg studied the stable carbon and nitrogen isotopes in the bone collagen (the main nonmineral component in bone). Values for carbon isotopes, in this case, will distinguish the extent to which the community was relying on Old World grains like wheat and oats, versus

BOX 2.1 Estimation of age from dental development

The development of a tooth is a marvelous thing (Hillson, 2014). Creating a shape that is complex yet consistent with its kind, various cells lay down material in a predictable way from inception through to young adulthood. Ameloblasts create enamel, odontoblasts create dentin. Whether the tooth bud is destined to be an incisor, canine or molar, the ameloblast will be first off the mark. The crown begins to take shape, to be lined with dentine. As growth progresses from what will become the chewing (occlusal) surface toward the root tip (apex), the tissues undergo structural alignment and maturation. It is only as a tooth emerges into the mouth that the root forms. As the deciduous teeth appear in a small child's mouth, the crowns of the permanent teeth develop behind each one. It is the growing permanent tooth crown that pushes out the baby tooth, causing the deciduous root to resorb so that the tooth is evulsed (and the tooth fairy is alerted). As it emerges, the permanent tooth has room to develop its root. The root is not completed for some time (Fig. B2.1).

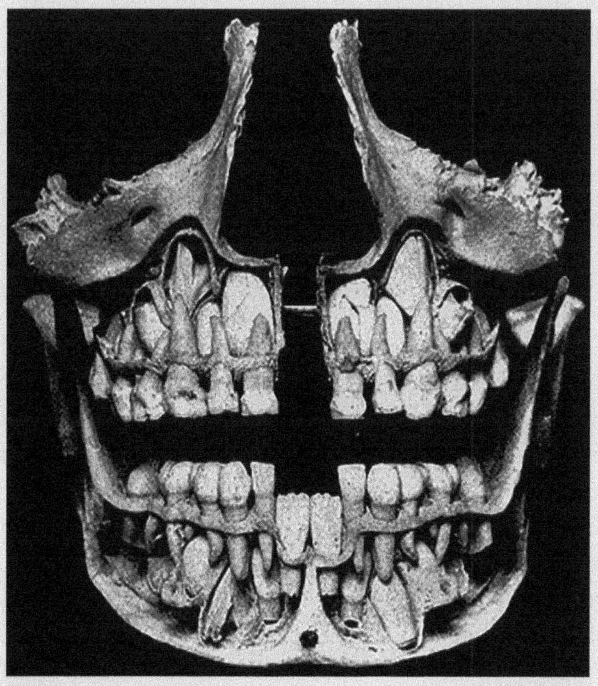

Fig. B2.1 A cut-away view of the jaws of a child of about 6.5 years, showing the emerged deciduous teeth and the permanent teeth to follow. *(Credit: Wikipedia "Deciduous teeth," accessed 02/16/2021.)*

Continued

BOX 2.1 Estimation of age from dental development—cont'd

The timing of these coordinated events within the jaws is relatively narrow in range. In circumstances of poor diet, febrile disease and other environmental stressors, the cells will continue their work, but will form tissue of compromised quality. From a bioarchaeological perspective, there are three important aspects of dental development:

1. The developmental status of a tooth is a reliable indicator of age-at-death, when skeletal remains are of an infant or child. Standards based on large and diverse samples of developing teeth (from x-rays) can be used to estimate an age range for an observed pattern of dental development (AlQahtani, Hector, & Liversidge, 2010, 2014).
2. The presence of an enamel defect (called hypoplasia because they signal a tissue deficiency) is a reliable, although nonspecific, indicator of environmental stresses that were encountered and overcome.
3. The elements and molecules that are recruited and captured as a tooth is built are reliable indicators of the growing child's diet and environment.

AlQahtani, S. J., Hector, M. P., & Liversidge, H. M. (2010). Brief communication: The London Atlas of human tooth development and eruption. *American Journal of Physical Anthropology, 142*, 481–490.

AlQahtani, S. J., Hector, M. P., & Liversidge, H. M. (2014). Accuracy of dental age estimation charts: Schour and Massler, Ubelaker, and the London Atlas. *American Journal of Physical Anthropology, 154*, 70–78.

Hillson, S. (2014). *Tooth Development in Human Evolution and Bioarchaeology*. Cambridge, U.K.: Cambridge University Press.

New World maize and cane sugar. The ratio of ^{15}N to ^{14}N gives an indication of the protein sources in the person's diet. Foods from higher on the food chain will be more enriched in ^{15}N than foods further down. Because suckling infants are feeding off their mothers' bodies, infants' bone that is formed during the breastfeeding period will be more enriched than the bone formed after weaning. By plotting the nitrogen isotope ratio against the age at death of each skeleton, we can estimate the age at which weaning was typically initiated. This study showed us several interesting things. The carbon isotope values of the whole sample indicate that people relied on wheat and oats for their bread and for their animals' fodder. They were not relying on maize, which was grown in the immediate vicinity by Indigenous communities. When compared to adult values, enriched nitrogen isotope values from the infants show that most were breastfed, with weaning begun just prior to their first birthdays and completed quickly. This is consistent with British historic documents for the era, which recommended initiation of weaning when the infant was between 8 and 12 months of age (Katzenberg & Pfeiffer, 1995). When I compare these results to those from other times and places, the mothers of Prospect Hill were remarkably similar to one another in their breastfeeding decisions. Usually there is a lot of diversity in mothers' approaches—when to start, how quickly to proceed—but not here.

Synopsis

And what about the young pregnant woman, the only person to have been autopsied? She was a delicately built young woman, about 157 cm tall, aged in her late 20s or early 30s. Her permanent third molars (wisdom teeth) were unerupted, tightly nestled horizontally in their sockets with nubbin-shaped roots: impacted. The bone of the roof of her eye sockets is pitted in a way that signals anemia, although we cannot say what caused it. The fetus was about 6–7 months gestation, and the tiny bones were normally formed. Was this a first pregnancy? There are some changes that can appear on surfaces of the anterior pelvic bones that are associated with the intense strain on connective tissue during childbirth. It would have been useful to assess those bones in this case, but their condition was too fragmentary. After death, the calvarium—the rounded part of the skull that surrounds the brain—had been sawed by a skillful surgeon, using manual tools. The grooves made by the saw teeth are clear. Autopsies were uncommon in this era. Normally, an autopsy would include the opening of the torso, done by removing the breastplate. That action slices some of the ribs, and sometimes the breastplate is discarded, thus missing from the buried body. The torso of this young woman was intact. Why did they examine the brain, but not the organs of the torso? They could weigh the brain; they could look for gross abnormality and asymmetry. Perhaps, if the woman had experienced a toxic reaction to the pregnancy, she might have been experiencing seizures prior to her death. The autopsy may represent a futile search for answers by those dealing with their grief. We do not know her name.

What did we learn about this little group of nineteenth century Euro-Canadians? We learned about their funeral customs and their child-rearing practices. We learned that they adhered to food traditions from their ancestral homeland, and that toothaches were probably common. The project illustrates the opportunities and challenges of cemetery excavations. A cemetery is an opportunistic sample, so we cannot attribute broad meaning to the proportions of age groups nor can we generate statistics about mortality rates. The requirement to do all of our initial work on site made us focus on only a few very specific questions. Because of the precise dental age estimates, those data and the bits of tissue have been useful to other researchers who are interested in normal child growth and the secular trend. That's the global tendency of people to get taller and mature earlier each decade. The discovery of the burials and the challenges of mitigation are typical of hundreds of such projects in many parts of the world. The expedient, unceremonious re-burial of the bones from Prospect Hill is not unusual. Discovery of human bones can be frustrating, costly and irritating to those involved.

Snake Hill

Occasionally, a discovery can be more than inconvenient: It can cause an international incident. As usual, this discovery began with the digging of a hole. In this case it was a basement for a new home that would join a row of homes along the shore of Lake Erie, on the Canadian side of the north-flowing Niagara River. Lake Erie drains into Lake Ontario via Niagara Falls. Less than twenty miles south of the falls is the source of the river, with Buffalo, New York on one side and Fort Erie, Ontario on the other. Modern commerce relies on the very busy Peace Bridge to cross the river. Things were not always peaceful along this waterway. The Niagara Frontier was a hot spot for hostilities during the War of 1812—an inconclusive war that both the Americans and the British feel they won. For a few months of 1814 American forces held Canada—which was at that time a British colony.

The excavation

At initial discovery, April 1987, attention was on the bones. They were buried rather shallowly and were clearly human. They might have been from long-ago foul play, an Indigenous peoples' cemetery, or an unrecorded family burial plot created by early settlers. Archeologists were engaged—the same company that excavated Prospect Hill, Archeological Services Inc. All musings about the origin of the bones abruptly ceased when a metal button was uncovered that was clearly from an American military uniform. That news led to what locals call "the second American invasion of Fort Erie" (Litt, Williamson, & Whitehorne, 1993, p. 31) by the U.S. Army Casualty and Mortuary Affairs Operations Division. This unit is dedicated, in part, to bringing home American soldiers who have died on foreign soil and doing so in a manner that conveys the honors they are due. International negotiations ensued. An international team was created to remove, assess, describe and repatriate the fallen soldiers to their home soil.

The military unit conveyed great urgency to remove the remains and winter was coming, so three large tents with massive heaters were installed. It was surprising that the burials were undisturbed, since a cottage and its septic system had sat above them since the 1940s. By December, the dig had uncovered the skeletons—full and partial—of thirty-one men, plus one ox skeleton, 440 buttons (5 British!), some straight pins probably associated with bandages, a handful of musket-shot, a pewter spoon and a wisp of hair, and three medical waste pits. The waste pits held the bones that had been surgically removed on site. Like at Prospect Hill, the buried bodies were on their backs, heads to the west, with hands folded over the pelvis, although here there were no coffins. The soldiers' skeletons showed dramatic evidence of their causes of death. One was missing his entire left leg. Another had a chunk of brick lying next to his spine. Some burials included extra appendages; other bodies were missing parts of limbs. There was even evidence of "friendly fire." The musket shot that had destroyed one man's face, still inside his skull, was American. One feature of any military occupation in that era was a surgery—a place to attend to the sick and wounded. The surgeon's craft in those days focused on simplifying complex traumatic wounds through amputation, without benefit of effective anesthetic or antibiotics. The waste pits held three amputated legs and two amputated arms. The arms, right and left, removed above the elbows, were almost certainly from the same person. His bones had still been growing (Fig. 2.4). He was a teenager.

Linking the bones with history

Both upper arm bones (humeri) had been sawn through at their narrowest points to simplify the wounds, removing the traumatically damaged forearms and hands. This would have made the intact upper arms easier to dress, it would minimize further blood loss and reduce the likelihood of infection. The medical waste pits provided my only experience of studying the skeletal remains of a people who didn't die—at least not just then. In that era, 40–60% of the soldiers who underwent amputation survived. The historian on the project team, Lt. Col. Joseph Whitehorne, was an expert in a field that was new to me: military history, complete with its memorabilia collectors, battleground visitors and re-enactors. He directed me toward various historical records of drummer boys, younger than the minimum age for recruitment (which was 14 years in that era). Their drums

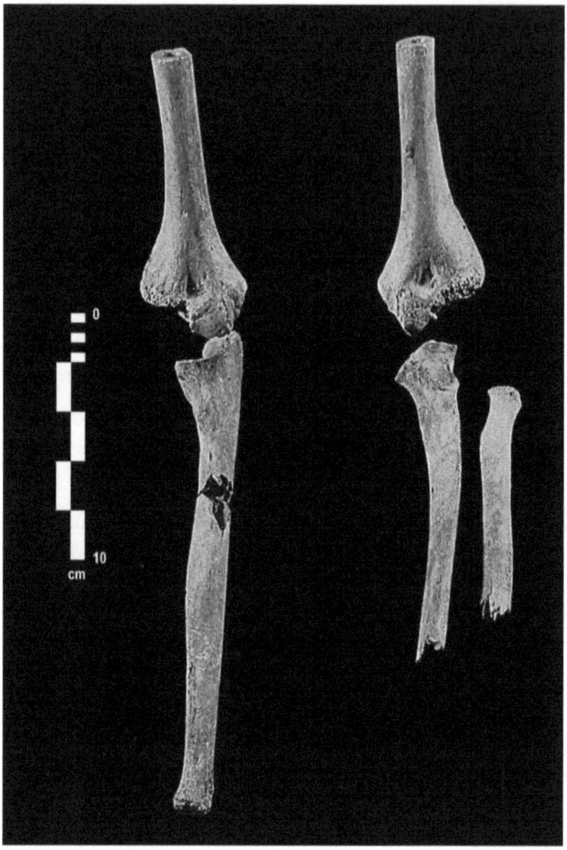

Fig. 2.4 The right and left elbow regions of a youth whose arms were amputated following trauma to the hand-wrist regions; found in a medical waste pit at Snake Hill, associated with the American occupation of Fort Erie, Ontario, during the War of 1812. *(Credit: Archaeological Services Inc.)*

communicated signals to the troops. They were regularly in harm's way. From the Revolutionary War through to the Civil War era, there are accounts of trauma experienced by these lads. I like to think that the lad who lost his arms at Snake Hill got back home and lived a long life as a proud veteran.

These were not the only youthful bones found at Snake Hill. Many of the more complete skeletons had not finished their maturation. Their youth was apparent in our study of their vertebrae—the components of the spine. At that time, the field commander Brigadier General Winfield Scott was known as one of the U.S. Army's great trainers. He prepared his troops through many hours of drill each day, during which they carried substantial equipment (Whitehorne, 1991). The Americans had taken Fort Erie in early July 1814, with few casualties on either side. They proceeded north, and laid siege to Fort George (their ultimate target). The two sides fought a major battle at Lundy's Lane on July 25. The Americans then withdrew to a reinforced Fort Erie in late July (they had left the Army Corps of Engineers to fortify the fort) with the British following and appearing in early August. The British then laid siege to Fort Erie. During August, both sides sought to advance their positions. The British abandoned the siege in September. In early November, the Americans blew up the fort and returned to the US.

Danger and hard work were the hallmarks of the soldiers' lives. It is not surprising that the spines of most men's skeletons showed the bone changes associated with herniated disks (Owsley, Mann, & Murphy, 1991). A well defined impression on the superior or inferior surface of a vertebral body, known as a cartilaginous (or Schmorl's) depression, is created when trauma causes fluid to escape the disk and press against the bone. The depressions normally occur on the bodies of vertebrae of the lower spine. While the correlation between bone changes and the experience of pain is not perfect, it is very likely that many of these young men "soldiered on" with significant back aches. One of the final changes that marks the full maturation of the human skeleton is the fusion of bony rings to the superior and inferior surfaces of each vertebral body. This happens at an age in the early 20s (Scheuer, Black, & Christie, 2000). The presence of Schmorl's depressions on vertebrae without fused rings is an illustration of the difficult lives of very young soldiers (Fig. 2.5).

We were given six months for study and analysis, after which the skeletons that could be associated with U.S. forces would be reburied. Twenty-eight complete skeletons could be identified as part of the American contingent with confidence enough to support repatriation. Three partial bodies and the amputated limbs could not be identified as American, and so were sent for curation to institutions dedicated to medical history in the two countries. The June 1988 repatriation ceremony for 28 flag-draped coffins holding the remains of American soldiers was a major international event. The carefully planned protocol included period re-enactors and an American military jet flyover. For the first and only time, the Peace Bridge was closed to traffic. After the Star-Spangled Banner, a nineteen-gun salute, an address, an invocation, The Last Post, one minute of silence, and Reveille, a lone bagpiper led a slow parade of 28 hearses over the bridge, heading toward interment at the military cemetery in Bath, New York (Litt et al., 1993).

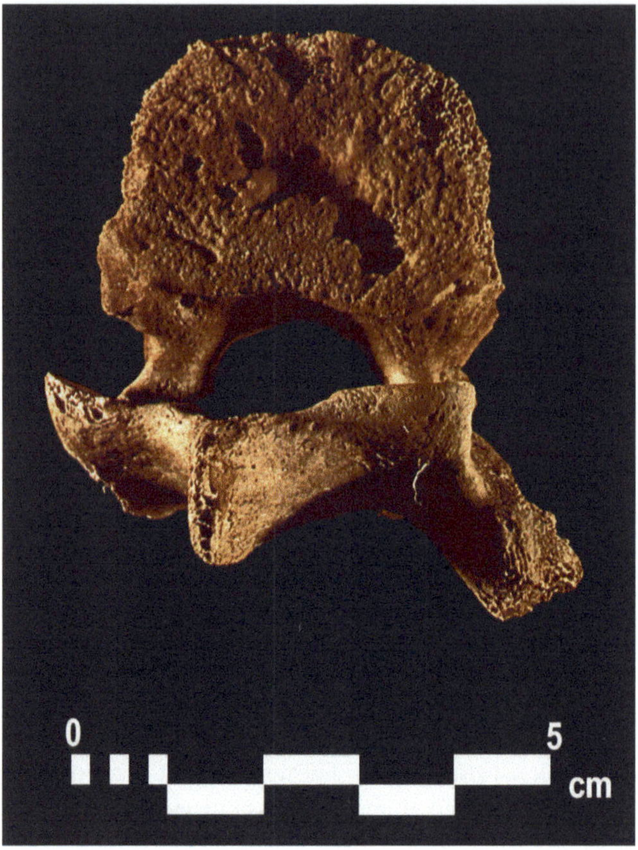

Fig. 2.5 A Schmorl's depression on the inferior body of a ninth thoracic vertebra, from a young soldier buried at Snake Hill. *(Credit: Archaeological Services Inc.)*

Two books (Litt et al., 1993; Pfeiffer & Williamson, 1991) were written about the excavation and its context. The first to appear (Pfeiffer & Williamson, 1991) is a compilation of reports by research team members. The chapters describe the injuries, the surgical approaches, and results from studies using traditional ways to figure out who these people were, like cranial shape and dental features. They also describe some approaches that were innovative at that time, like measures of lead exposure and use of stable isotopes to associate a skeleton with its place of origin. The research team was focused on establishing identities, but we could not get more specific than attribution of nationality, and that was most confidently based on buttons, when they could be associated with bones. Despite access to some documentary information about the 2800 soldiers who served on the Niagara Frontier, no individual could be given a name, although the list was narrowed to about 100 candidates. Thanks to subsequent scientific advances, perhaps palaeogenomics could now be applied. Each reburied skeleton is identified by number, just in case someone wants to revisit the identity question.

The second book (Litt et al., 1993) describes the discovery and historical context that I have summarized very briefly here. It has proved popular with readers who are keen on military history. What a remarkable few months they had been. Why did I devote time to Snake Hill here? It is an example of a high profile, well supported excavation, analysis, repatriation and reburial. Because of the U.S. Army's mandate, once the bones were discovered, time for study was tightly constrained. Indeed, plans for the reburial were underway before the excavation was complete. Despite the absence of any personal identification, these unknown soldiers were honored because of their status as soldiers and their actions in life, both documented and imagined. They had died on foreign soil. This fact catalyzed American laws and customs. Many Americans assume that all nations have similar practices. Some do, most don't.

Historic graves in broader context

These stories illustrate that even relatively recent in-ground burials are sometimes disturbed. In such instances, disposition of the human remains can present economic and regulatory dilemmas. It is sometimes argued that European-origin North Americans would never think of treating their ancestors' remains in anything but the most respectful fashion. That is just not so. Even for people with the best of intentions, human skeletons sometimes cause inconvenience, expense and consternation. The treatment of the unnamed dead is contingent on economics, legislation, and culture. Although there are normally legal parameters to be followed, there is no single approach applied to the disposition of human remains, once they have been discovered.

In the chapters that follow, I describe research into the more distant past, in which the human remains are not European (sometimes termed "Western" or "First World") in their ethnicity. While there are many parallels, there is also one new factor, that of cultural dissonance. The people doing the excavation and research are, for the most part, of European origin. They are studying the remains of people who are not their ancestors, except in the broad sense of collective humanity. Histories of conflict between groups and colonialist attitudes affect the perspectives of everyone involved. It is not difficult to see this as a power imbalance. One of the themes of this book is how to address it.

References

Katzenberg, M. A., & Pfeiffer, S. (1995). Nitrogen isotope evidence for weaning age in a nineteenth century Canadian skeletal sample. In A. L. Grauer (Ed.), *Bodies of evidence: Reconstructing history through skeletal analysis* (pp. 221–236). NY: Wiley-Liss, Inc.

Litt, P., Williamson, R. F., & Whitehorne, J. W. A. (1993). *Death at Snake Hill: Secrets from a war of 1812 cemetery*. Toronto: Dundurn Press.

Owsley, D. W., Mann, R. W., & Murphy, S. P. (1991). Injuries, surgical care and disease. In S. Pfeiffer, & R. F. Williamson (Eds.), *Snake Hill: An investigation of a military cemetery from the war of 1812* (pp. 198–226). Toronto: Dundurn Press.

Pfeiffer, S., Dudar, J. C., & Austin, S. (1989). Prospect Hill: Skeletal remains from a 19th century Methodist cemetery, Newmarket, Ontario. *Northeast Historical Archaeology, 18*, 29–48.

Pfeiffer, S., & Williamson, R. F. (Eds.). (1991). *Snake Hill: An investigation of a military cemetery from the war of 1812*. Toronto: Dundurn Press.

Saunders, S. R., & Lazenby, R. (Eds.). (1991). *The links that bind: The Harvie family nineteeth century burying ground*. Dundas, Ontario: Copetown Press.

Scheuer, L., Black, S., & Christie, A. (2000). *Developmental Juvenile Osteology*. San Diego: Elsevier Ltd.

Whitehorne, J. W. A. (1991). Fort Erie and U.S. operations on the Niagara frontier, 1814. In S. Pfeiffer, & R. F. Williamson (Eds.), *Snake Hill: An investigation of a military cemetery from the war of 1812* (pp. 25–60). Toronto: Dundurn Press.

Whiteley, M. F., Pfeiffer, S., Austin, S., & Dudar, J. C. (1993). Archaeology, biology and history: Research collaboration in the Prospect Hill project. *Papers of the Canadian Methodist Historical Society, 9*, 65–77.

CHAPTER 3

Framework for Osteobiographies: Lower Great Lakes

Prologue

Changes were planned. Soon we would move the boxes that held the bones and teeth that had been unearthed by the university's archeological excavations. The boxes would be taken to a building where the preparations for reburial would take place. But first, descendants would pay their respects. A consortium of First Nations groups had asked to perform a cleansing ceremony. An important part was to be the burning of a sweetgrass braid. Participants would gesture so that the smoke washed over their bodies in a spirit of kindness and compassion, to heal the mind and the spirit (Kimmerer, 2013). There was a bit of inconvenience to the university, since the building's smoke detectors had to be de-activated during the appointed time. That had been arranged. What was more difficult to accommodate was the feeling of awkward embarrassment that the boxes—with their dissimilar sizes and markings—were stacked high in a storage room, in a manner that the tribes' representatives would likely judge to be disrespectful. We had been getting to know one another as we sat around meeting tables for many months. I hoped that the First Nations representatives had at least tentatively concluded that our good intentions aligned with theirs, but the religious leaders who would perform the ceremony had not been to campus before. What would they think of us?

The day arrived, and we anthropologists felt ready to receive our visitors. We would present each representative with a pouch of tobacco and we would share food with them after the ceremony. Traditional religious leaders from Anishnaabe, Haudenosaunee and Huron-Wendat communities arrived, and they began the smudge. Within the room that held their ancestors' bones, the three men stood close, facing one another, around the shell holding the smoking fire. The scent of sweetgrass filled the space, carrying with it our thoughts and feelings. We each drew the smoke over ourselves to cleanse ourselves. The three leaders spoke in turn, each addressing the four directions. I do not have a strong ear for languages, but I could discern

Osteobiographies
https://doi.org/10.1016/B978-0-12-823880-6.00004-5

the distinctiveness of the Anishnaabe words (an Algonquian language). It seemed to me that the Haudenosaunee and Huron-Wendat presentations sounded linguistically similar, but I reasoned that since both are Northern Iroquoian groups their languages would be related. The atmosphere was calming, with a pervasive sense of cleansing. I felt great appreciation toward all the people who had collaborated to make it happen. After the ceremony, our First Nations friends explained its content: The spirits of those many souls whose journeys to the Land of the Dead had been disrupted had been comforted by re-connection with people who understood their unease, people who would do their best to bring them to a resting place from which their journey could continue. My unasked question about language was answered by a Haudenosaunee participant. "Did you notice that our Huron-Wendat friend spoke mostly in Haudenosaunee? That's because most of their language has been lost." I have learned since that the Huron-Wendat Nation has made progress to rediscover and establish their language.

Hundreds of distinctive cultures existed in the Americas prior to the disruption of European arrivals. Just like everywhere else in the world, these cultures had distinctive languages, spiritual beliefs, strategies for feeding and clothing themselves, ways of raising their children and having fun. All of these cultures took a blow when new people, new diseases, and new technology abruptly appeared in the early 17th century. The Wendat Nation, successfully farming, fishing and hunting in the region that the newcomer Europeans called Upper Canada, lived in villages spread across a broad territory. Their name for themselves, "Wendat," means "islanders" or "dwellers on a peninsula" as their core territory was a peninsula extending into Georgian Bay, on Lake Huron, one of the Great Lakes (Williamson, 2016). The French referred to them as *les Hurone*, a term from Old French term meaning "shaggy or unkempt." Today, the geographic region is often referred to as "Huronia". The Huron-Wendat people know it as Old Wendake (Fig. 3.1). Diseases like measles and smallpox, exacerbated by warfare, reduced the Nation to several hundred when survivors fled their homeland in 1649–51 CE. Thanks to the combined sources of some surviving oral tradition, documents from European visitors and archaeology, we know quite a bit about their world.

History's story line

Like on-line posts from modern travelers, the early reports sent back to Quebec (ultimately to France) were selective descriptions. They reflect the features important to the writers' jobs as explorers, traders and missionaries.

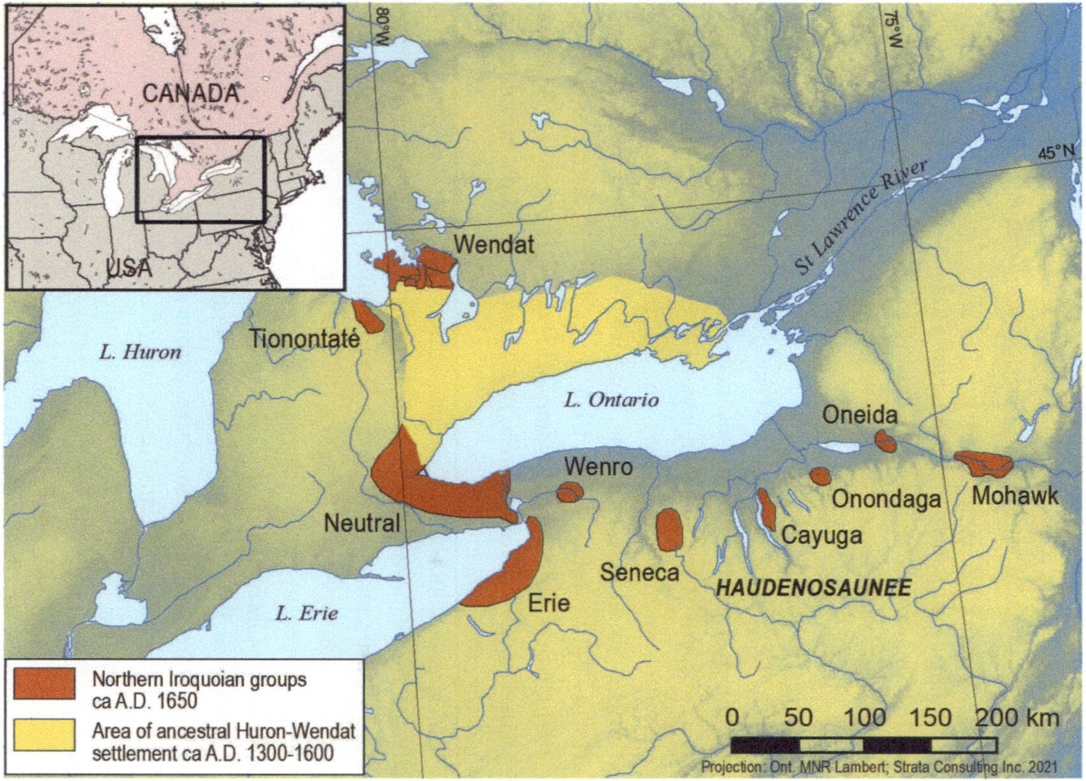

Fig. 3.1 Northern Iroquoian-speaking groups, approximately 1300–1600 CE (courtesy of Ron Williamson).

The Wendat, because of their location and their political network, were able to supply the French with vast quantities of beaver pelts and other valued natural resources. In exchange, they received European kettles, knives, hatchets, and other goods. The Wendat also provided food and shelter to intermittently visiting newcomers for about 35 years. The first (that we know) was an emissary of the explorer Champlain: Étienne Brûlé. In 1610 CE, he was sent to live with an Algonquin band who spent their winter with the Huron. He was instructed to learn the Wendat language. The explorer Samuel de Champlain spent the winter of 1615–16 CE in Huronia. The Recollect friar Gabriel Sagard visited in 1623–24, and representatives of the Jesuit Order began intensive work among the Huron in 1634. English translation of their annual reports was first accomplished by R.G. Thwaites and associates, who published 73 volumes during the years 1896–1901 (known as "The Jesuit Relations"). Many documents survive. One of the most touching is the "Huron Christmas Carol," written in the Huron-Wendat language by a Jesuit Father, Jean de Brébeuf. It is one of a very few language remnants. Modern carolers sing the English version, published as "T'was in the Moon of Wintertime," and it is lovely. A number of excellent scholarly books (Heidenreich, 1971; Sioui, 1999; Tooker, 1964; Trigger, 1969, 1976), works of fiction (Boyden, 2013) and a few movies (Beresford, 1991; Dansereau, 1965; Sioui Labelle, 1999) have been produced, describing and imagining this dynamic period.

What happened in 1649 CE? Competition among Indigenous groups for access to the lucrative trade with Europeans had gradually transformed traditional rivalries into serious warfare. The Iroquois (Haudenosaunee) had more access to guns. By the spring of 1649, many Wendat had been killed, captured or forced to flee by the Haudenosaunee. About three hundred Huron-Wendat Christians spent the winter of 1649–50 living on an island in Georgian Bay, then moved, under French protection, to a locale in Quebec. Others joined them there within a year or so. That community, Wendake, is now the Canadian home of the Huron-Wendat Nation, numbering about 3000 people. Other descendants of the dispersed Nation formed the Wyandot people of Kansas, Michigan and Oklahoma, numbering about 6000, and a nonstatus group based in Ontario who live near Detroit, the Wyandot of Anderdon Nation.

The historical documents are of great value, but they reflect the perspectives of the writers. As Bruce Trigger (the premier scholar of all things Huron-Wendat) observed, "In this exotic cultural environment, individual Frenchmen found themselves alternately scandalized, repelled, and fascinated by the behavior of a people who were uninfluenced by beliefs and standards they themselves had never questioned" (Trigger, 1969:2). Only a few communities hosted the visitors, and many activities of daily living were of little interest to the diarists. The presence of the traders and missionaries, their foreign goods and their destabilizing interventions all influenced the culture being observed. The nascent antagonisms between Indigenous regional groups were escalating, thanks to economic incentives and the Europeans' lethal hardware. Some very interesting archeological investigations focus on linking evidence from Huron-Wendat sites to events recorded by the French, rather like the way Biblical archeologists try to link material evidence to that document. We can learn from all this, but it cannot take us further back than the turn of the seventeenth century.

Archaeology's story line

Huron-Wendat culture as it existed when the Europeans arrived can be traced back to roots hundreds of years earlier. This is where the archeological record really kicks in. In the absence of written documents, the archeological evidence—the locations and structure of sites and artifacts, the human remains—provides crucial testament to when and how a culture developed through time.

The decision about where to begin this story is arbitrary. Rather than beginning with the migration of the first humans onto the North American continent, the human story for this area (Fig. 3.1) begins when glacial waters receded from what is now the Great Lakes region. The earliest evidence for people on the broad landscape that we now know as southern Ontario dates to about 13,000 years ago. Vestiges of these early hunters are rare. They were following the game animals on which they relied. A slightly better record exists for the hunter-gatherers who exploited the region's plants, animals and waterways from about 6000 years ago. This is the mid-Holocene in geological terms (Munson & Jamieson, 2013). To get an idea of how scant the record is, consider that the oldest evidence of a dwelling is from only about 3000 years ago (Ellis, Keron, Menzies, Monckton, & Stewart, 2015). My PhD research was on the human remains from that time period. These hunter-gatherers probably moved seasonally, but they returned to specific places to bury their family members. I will write more about them later (Chapter 4).

Traditionally, archeologists have applied general names to the different periods. The big game hunters are known as Paleo-Indians and the hunter-gatherers are from the Archaic period. Such names are short forms for cultures that were more complex and multifaceted than the simple labels imply. The term "Archaic" can be especially confusing, since the word has many other general applications in English. Even more confusing is the period that comes next. About 2800 years ago began the Woodland period, marked by the introduction of domesticated plants, and progressively greater reliance on them. This is not to say that there were no woodlands in earlier times! There were plenty of trees, forming a well-watered deciduous forest that provided nuts, hardwood for tools and softwoods for fuel, interspersed with meadowlands and wetlands. Archaic peoples had been relying on the woodlands for millennia. But once people began systematically growing their food, we call it the Woodland period—divided into Early, Middle and Late.

The domestication of *Zea maize*, known to North Americans as corn, began in Mexico, about 9000 years ago. Why did it take so long to reach the Great Lakes? Because the Great Lakes growing season is quite short, compared to more southerly locations. As it was traded up the Mississippi and Ohio River drainages, Indigenous peoples selectively bred the seed for their local climates (Monckton, 2013). Other plants, like beans, squash and tobacco were also in play, all coming from the south. On the other hand, goosefoot (a form of quinoa) is a local weed that may have been domesticated long before the exotics arrived (Crawford, Lytle, Williamson, & Wojtowicz, 2018). By about 1000 years ago, some people were relatively settled, relying in part on food they planted themselves. They are the Woodland people.

When group members died

Thousands of people, over thousands of years, lived and died in this region. When group members passed, each community had ways of responding that were both practical and symbolic. These traditions were distinctive and sustained, so that they can be used as cultural identifiers. The practices of people who lived in what is now southern Ontario are diverse. If one were to consider broader swathes of North America, there would be even more diversity. The similarities of practice—in North America and around the world—are found in the practicalities and the symbolism of connection and continuity, not in the specific activities.

The Archaic hunter-gatherers buried their dead at special locales (Walker, 2019). Located at a spot both rich in food resources and sandy for ease of digging, it appears that each cemetery was used for many years by one or more groups who visited the spot, perhaps in an annual cycle. Some of the burials are of thoroughly burnt bone, others are of bodies buried soon after death. This may indicate that those who died when the group was far away were cremated so that the remains could be transported to the family plot. Sometimes a cremated burial feature contains the remains of more than one person. Of those buried with the body intact, they are usually tightly flexed on their sides. Sometimes there is more than one person in a grave. Positioning is irregular enough to suggest that the specific locations of prior burials were not clearly marked on the surface. Whether cremated or not, functional and decorative objects as well as food were often placed with the deceased (Donaldson & Wortner, 1995).

The approach to honoring the dead followed by the Woodland people we identify as ancestral Huron-Wendat, many generations later, is both different from and similar to that of Archaic people. Like the hunter-gatherers, there were specific traditions to be followed, with preparation of the deceased for the next world evident through the presence of artifacts and food. The difference is that, for the most part, the ancestral Huron-Wendat followed the initial primary burial of the deceased with a collective secondary burial, creating a community ossuary. Through both history and archaeology, there is an extensive literature on when, how, and why Huron-Wendat ossuaries were created (Robertson, 2004; Seeman, 2011; Tooker, 1964; Williamson & Steiss, 2003). Wendat ossuaries are large, rather deep, pit features that contain the disarticulated and commingled remains of individuals who were interred during an important ceremony known as the "Feast of the Dead." One such ceremony was witnessed in 1636 CE by Jean de Brébeuf in historic Wendake (Thwaites, 1896–1901, 10:279–303) (Fig. 3.2). At a time determined by the decision to relocate a village, the remains of community members who had died—whose remains had been initially buried or stored—were placed into the collective grave. The number of persons interred ranges from fewer than 100 to several hundred (Williamson & Steiss, 2003). It is tricky to extrapolate from what was observed by European visitors back to the earlier, precontact days, but archeological evidence suggests that this pattern had been in place for at least four hundred years.

The contemporary context

While the region explored here is only a fraction of the North American landscape, it serves to illustrate some broader patterns. An ancestor-descendant relationship can be asserted for the Huron-Wendat, while that

Fig. 3.2 An engraving that depicts the Feast of the Dead, based on Brébeuf's account (J.-F. Lafitau, Moeurs des sauvages amêriquains, II, 1724). (Source: Wikipedia "The Huron Feast of the Dead" accessed 02/19/21.)

kind of link between past and present is more tenuous for the Archaic hunter-gatherers. Nevertheless, all excavated human remains can be seen as ancestral, in a general sense, to living communities who trace their ancestry to Indigenous roots. As noted earlier in this chapter, the Huron-Wendat have been actively engaged with repatriating their ancestors. For them, the present and the past blend seamlessly. Clearly, anyone who was buried within a Huron-Wendat ossuary is an ancestor to the contemporary descendants. This inevitably draws attention to the collections of bones in boxes curated by museums and universities and brings in the topic of government policies. While the focus here is the Canadian province of Ontario, the expectations of the populace are similarly reflected through governance in other jurisdictions (Box 3.1).

Box 3.1 A lexicon of those who study death

Different human communities can feel very strongly about the manner in which a deceased person is regarded and handled. Laws prohibiting desecration of the body or a cemetery are common around the world, with each jurisdiction having a pretty firm sense of what is proper behavior and what is not. Yet, if one scans practices through space and time, the approaches considered right and proper vary greatly. It is not surprising that anthropologists and archeologists study these practices, as a part of their attempt to understand humanity through the diversity of their preferred practices through space and time.

A central distinction is that bioarchaeologists focus on learning about past people from their physical remains, while various other approaches focus on cultural responses to death. To understand life and death in the past, we need both sides of the story.

Bioarchaeology/osteoarchaeology focus on the human remains, in their archeological context (Martin, Harrod, & Pérez, 2013). The goal is to uncover aspects of the lives lived by the deceased. Bioarchaeologists interact with workers who take other perspectives, like the following:

Funerary/Burial/Mortuary Archaeology focus on practices related to the disposal of the dead (Bradbury & Scarre, 2017), such as funeral rites and burial. In modern usage "mortuary" applies to the preparation of the body and "funerary" applies to the deposition of the body. However, in practice these terms are interchanged by researchers studying past peoples. The meaning of the word "mortuary" is "pertaining to death," so it is the broader term. Various postgraduate programs are titled with one or more of these terms, incorporating varying ratios of archaeology and bioarchaeology/osteoarchaeology. Their content expands into cemetery archaeology.

Continued

Box 3.1 A lexicon of those who study death—cont'd

Cemetery Archaeology focuses on the study of designated burial grounds, through their identification, mapping and documentation.

Death Studies is a relatively new focus among archeologists. It is meant to combine the fields listed above with input from textual specialists including historians and theologists, to broaden the interpretive scope. However, the term "death studies" is also broadly used for explorations of contemporary death, dying, bereavement and death education. These are the topics of interest in a journal by that name. Use of this term by researchers of the distant past may not be workable, despite its attractive scope.

Forensic anthropology and forensic archaeology: The term "forensic" means "belonging to the courts," or of legal relevance. The theory and methods of these rapidly developing fields have provided nonforensic researchers with new perspectives and approaches. Forensic researchers have developed and adopted technological and methodological innovations. In most parts of the world, when human remains are initially discovered, a determination is made by a coroner or medical examiner regarding forensic relevance. Jurisdictions identify—for legal purposes—a calendar date beyond which the legal framework is not applicable. If remains are deemed to date from a time prior to the cut-off, they are treated as archeological discoveries.

Bradbury, J., & Scarre, C. (Eds.). (2017). *Engaging with the Dead: Exploring Changing Human Beliefs about Death, Mortality and the Human Body* (Vol. 13). Oxford, U.K.: Oxbow Books.

Martin, D. L., Harrod, R. P., & Pérez, V. R. (2013). *Bioarchaeology: An Integrated Approach to Working with Human Remains.* New York: Springer-Verlag.

Early practicing archeologists were keenly interested in the structure of ossuaries and in the artifacts within them. As well, expanding urban development led to disturbance of unmarked cemeteries. The human remains from such archeological excavations undertaken prior to the establishment of The Ontario Heritage Act (first enacted in 1975) were held by institutions like museums and universities on behalf of the broader public. While a step up from the opportunistic looting for anatomical collections and race science research that occurred in the late nineteenth and early twentieth centuries (Williamson, 2014), their curation and use followed norms of the day which seem quite callous to modern eyes. These shifted with the decades. By the 1970s, the use of these collections for teaching was broadly deemed inappropriate and the collections were gradually seen less as a resource and more as a cause for concern. Meanwhile, government oversight sought to avoid disturbance of human remains except in situations of emergency or immediate threat of destruction, and the intent to consult with descendants gradually crystalized.

Since the 1970s, there have been many reburials of Indigenous ancestral remains. Whenever an emergency mitigation is undertaken, provincial government policies mandate that an Indigenous group be contacted and consulted about the disposition of the remains. Management of heritage is a mainly provincial responsibility in Canada, so Ontario first simply sought the input of the nearest provincially recognized Indigenous group. This could be awkward, since most of the remains encountered were of ancestral Huron-Wendat whose descendants live in the province of Quebec and the United States. The Huron-Wendat of Wendake, Quebec wished very much to be engaged with these encounters with their ancestors. They helped the province find a work-around, and since the 2005 Teston Road case described below, the Huron-Wendat of Quebec have been consulted.

Several ossuaries have been reburied soon after excavation. Late in 1997, the human remains and included materials (often called "grave goods") of the Moatfield ossuary (minimally 83 persons) were reburied very near where they had been removed during an emergency excavation. That collaboration was between Archeological Services Inc. (ASI Heritage) and the Six Nations of the Grand, a nation of Haudenosaunee people who were designated to play the descendant community role by the Ontario government (Williamson & Pfeiffer, 2003). In 2005, a widening of Teston Road, north of Toronto, exposed an ossuary (estimated to have held 300–400 persons), and the Mississaugas of Scugog Island First Nation, a nation of Anishnaabe people, were contacted first. The Huron-Wendat were subsequently consulted and were involved in the discussions about what to do with the site. In that case, agreement was reached to realign the road, rebury the disturbed remains there, cover them with a steel wire mesh, then secure the site with concrete. On the surface, a large rock was placed over the ossuary (R.F. Williamson, pers. com.).

Such consultations do not apply solely to ossuaries, of course. There have been many arrangements made, each unique to their circumstances of discovery, investigation, and governance contexts. In 2001, individual burials were partially exposed by erosion in a public park in the town of Windsor, Ontario. The positions of the bodies were analogous to typical placement in a Christian settlers' cemetery, but there was no sign of coffins. After partial exposure, they were radiocarbon dated to about 1000 years ago. The burials were linked to ancestral Indigenous people, some of whose funerary traditions mirrored those of the later settlers. The nearby Walpole Island First Nation worked with the city to arrange re-zoning of that area as a cemetery (Dewar, Ginter, Shook, Ferris, & Henderson, 2010). The examples from Ontario illustrate current practice in one provincial jurisdiction. There are many parallels in places throughout the world where there are frameworks for heritage oversight. Despite the attractive features of respect and consultation, this framework is silent on the important matter of what we can learn from human remains. Unless the archeologist requests consideration of some analysis, none will be done. From my perspective, this is a problem.

The collections that are held in institutions present a different set of issues, but consultation and reburial are happening there, too. Because curated human remains have been formally accessioned, each institution must grapple with its own policies when they weigh a request and potentially de-accession material. In 1999, the human remains and grave goods of the Ossossané ossuary (excavated 1947–48) were removed from the Royal Ontario Museum (ROM) and reinterred at their original place of burial. This was primarily a collaboration between the ROM and the Wyandot groups of the US (Kapches, 2010). In 2013, The University of Toronto transferred to the Huron-Wendat of Quebec control over the human remains and grave goods of twelve archeological sites. They then reburied their ancestors, the number estimated to be about 1700, at one of the original ossuary sites (Forrest, Williamson, Pfeiffer, & Lesage, 2020; Pfeiffer & Lesage, 2014). I was involved in the long, complex discussions that preceded this event. It forms the context for some of what I will describe in chapters 5 and 6.

Summary

The sector of North America known as the Lower Great Lakes has been home to humans for a brief time, as compared to much of the rest of the world—only a few thousand years. Once liberated from glacial ice and water, a landscape developed that supported humans who followed a mobile, hunting-gathering approach to survival. This links them, anthropologically-speaking, to hunter-gatherers in many other parts of the world. Understanding how they survived and flourished can provide insights into how humans everywhere survived prior to the changes that came with access to domesticated plants and establishment of more permanent residences. During the past thousand years, the Lower Great Lakes provided productive space for maize and other crops. We know more about the farmers who lived in the region subsequent to the hunter-gatherers, because the archeological vestiges are more apparent and because the farmers hosted the Europeans who arrived in the seventeenth century. This generated valuable written documents.

Learning about Indigenous ancestors can be equated to helping descendants learn about their family. The people who followed the two subsistence approaches that were briefly introduced here also created and followed very different burial practices. Archaic people created cemeteries of cremations and primary burials. Ancestral Huron-Wendat people created secondary shared burials known as ossuaries. The juxtaposition of these two lifeways, separated in time, provides an opportunity to see how past communities organized their lives. The study of their ancestral remains, as extracted from archeological contexts, can generate new knowledge about the ancestors. This in turn contributes to a broader understanding of human survival, and to a deeper understanding of distinctive cultures.

References

Williamson, R. F., & Pfeiffer, S. (Eds.). (2003). *Vol. 163. Bones of the ancestors: The archaeology and osteobiography of the Moatfield site*. Ottawa: Canadian Museum of Civilization.

Williamson, R. F., & Steiss, D. A. (2003). A history of Iroquoian multiple burial practice. In R. F. Williamson, & S. Pfeiffer (Eds.), *Bones of the ancestors: The archaeology and Osteobiography of the Moatfield ossuary* (pp. 89–132). Ottawa: Canadian Museum of Civilization.

Beresford, B. (Writer). (1991). Black robe. (In. Canada).

Boyden. (2013). *The Orenda*. Toronto: Hamish Hamilton.

Dansereau, F. (Writer). (1965). Mission of fear/Astataïon, ou Le Festin des morts. (In. Canada).

Crawford, G. W., Lytle, J. L., Williamson, R. F., & Wojtowicz, R. (2018). An Early Woodland domesticated chenopod (*Chenopodium berlandeieri* subsp. *Jonesianum*) cache from the Tutela Heights site, Ontario, Canada. *American Antiquity*, *1*. https://doi.org/10.1017/aaq.2018.75.

Dewar, G., Ginter, J. K., Shook, B. A. S., Ferris, N., & Henderson, H. (2010). A bioarchaeological study of a Western Basin tradition cemetery on the Detroit River. *Journal of Archaeological Science*, *37*(9), 2245–2254. https://doi.org/10.1016/j.jas.2010.03.022.

Donaldson, W. S., & Wortner, S. (1995). The Hind site and the glacial kame burial complex in Ontario. *Ontario Archaeology*, *59*, 5–95.

Ellis, C. J., Keron, J. R., Menzies, J., Monckton, S. G., & Stewart, A. (2015). For immediate occupancy: Cozy three-thousand-year-old heritage winter house with river view near Lake Huron—apply to Terminal Archaic realty. In B. G. Redmond, & R. A. Genheimer (Eds.), *Building the past: Prehistoric wooden post architecture in the Ohio Valley—Great Lakes* (pp. 29–62). Gainsville: University Press of Florida.

Forrest, C. L., Williamson, R. F., Pfeiffer, S., & Lesage, L. (2020). The joy of the souls: The return of the Huron-Wendat ancestors. In C. H. Meloche, L. Spake, & K. L. Nichols (Eds.), *Working with and for ancestors: Collaboration in the care and study of ancestral remains* (pp. 151–165). London and New York: Routledge.

Heidenreich, C. E. (1971). *Huronia: A History and Geography of the Huron Indians, 1600–1650*. Toronto: McClelland and Stewart Ltd.

Kapches, M. (2010). Ossossané ossuary: The circle closes. *Archaeology of Eastern North America, 38*, 1–15.

Kimmerer, R. W. (2013). *Braiding Sweetgrass: Indigenous wisdom, scientific knowledge and the teachings of plants*. Minneapolis: Milkweed Editions.

Monckton, S. G. (2013). Plants and the archaeology of the invisible. In M. K. Munson, & S. M. Jamieson (Eds.), *Before Ontario: The archaeology of a province* (pp. 124–133). Montreal & Kingston: McGill-Queen's University Press.

Munson, M. K., & Jamieson, S. M. (Eds.). (2013). *Before Ontario*. Montreal & Kingston: McGill Queen's University Press.

Pfeiffer, S., & Lesage, L. (2014). The repatriation of Wendat ancestors/Le rapatriement des ancêtres Wendat, 2013. *Canadian Journal of Archaeology, 38*, 1–22.

Robertson, D. A. (2004). The Hutchinson site: A place to prepare for the final journey. *Ontario Archaeology*, 77(78), 95–120.

Seeman, E. R. (2011). *The Huron-Wendat feast of the dead: Indian-European encounters in early North America*. Baltimore: Johns Hopkins University Press.

Sioui, G. E. (1999). *Huron-Wendat: The Heritage of the Circle*. English version ed. Vancouver: UBC Press.

Sioui Labelle, R. (Writer). (1999). Kanata: Legacy of the children of Aataentsic. In. Canada: National Film Board.

Tooker, E. (1964). *An Ethnography of the Huron Indians, 1615-1649*. Vol. Bulletin 190 Washington, D.C.: U.S. Government Printing Office.

Trigger, B. G. (1969). *The Huron farmers of the north*. New York: Holt, Rinehart and Winston.

Trigger, B. G. (1976). *The children of Aataentsic: A history of the Huron people to 1660*. Montreal & Kingston: McGill-Queen's University Press.

Walker, S. (2019). The persistence of place: Hunter-gatherer mortuary practices and land-use in the Trent Valley, Ontario. *Journal of Anthropological Archaeology, 54*, 133–148.

Williamson, R. F. (2014). The archaeological history of the Wendat to A.D. 1651: An overview. *Ontario Archaeology, 94*, 3–64.

Williamson, R. F. (2016). East-west interaction among fifteenth-century St. Lawrence Iroquoian and north shore of Lake Ontario ancestral Wendat communities. *Ontario Archaeology, 96*, 104–120.

CHAPTER 4

Mobile people making a go of it

Prologue

The landscape immediately north of Lake Erie is very flat. People tend to speed as they drive the multilane 250-mile road that connects Detroit and Toronto, unless it is winter and they get snared by the infamous "snow belt" around the city of London, Ontario (Fig. 4.1). It is an agricultural landscape. There are the pungently aromatic tomato fields around Leamington, known since 1908 as the Tomato Capital of Canada (Wikipedia). There is the tobacco belt around Tillsonburg, where high quality leaves are produced for the export market, and there are thousands of hectares of corn and soybeans. It is hard to imagine this landscape as it existed prior to an aggressive program of draining and tiling in the 19th century. When glacial waters receded, about 13,000 years ago, this was a landscape that attracted some of the last North American mastodons. As the landscape developed, the streams and wetlands supported a wide array of fish and other marine foods. There were areas of mixed forest that included nut- and fruit-bearing trees and bushes. Where there is food there will be people, and so people from the south migrated into the region, although evidence for the very earliest human presence is exceedingly scarce (Storck, 2004).

My experience with these farmlands began in January 1971, with the arrival of a message to a professor at University of Toronto from an avocational archeologist who farmed land near the village of Bothwell, Ontario. Since 1968, Stanley Wortner had meticulously documented the gradual exposure of human remains as winds eroded a sand knoll on a neighbor's property (Mr. Herman Hind, hence the "Hind Site," officially designated as AdHk-1 in the provincial log). Wortner's count was up to sixteen burials, and he wanted some help. As a newly admitted PhD student, I was keen. Thanks to Wortner's letter, the topic of my thesis research began to take shape.

When I arrived at Wortner's farm that spring, my first assigned task was to spend a day deployed on a tree planter, helping him plant a windbreak

Osteobiographies
https://doi.org/10.1016/B978-0-12-823880-6.00005-7

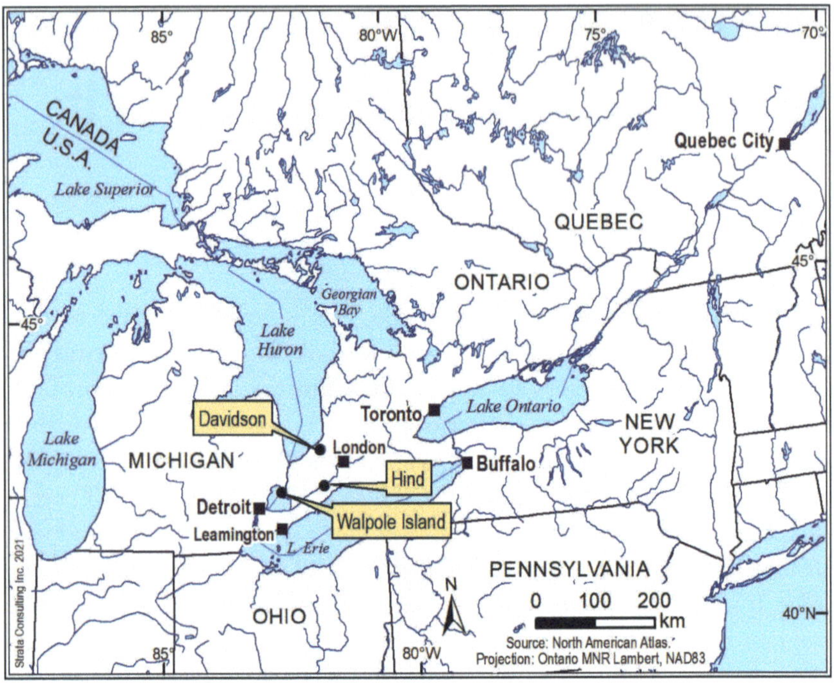

Fig. 4.1 A map of the region discussed in this chapter.

of walnut trees alongside the farmhouse. Once that was done, we could get to work in the nearby field, documenting the fascinating context of the cemetery area. Wortner surmised that its features suggested it had been established by hunter-gatherers of the Late (or Terminal) Archaic, dating to 4500–2800 years ago (Ellis, 2013). The site was a slightly raised sand bar created by an old meander of the nearby Thames River. The cremations and primary burials were in fine sand. When dry, they could be disturbed by the slightest bit of wind, so the excavators worked on the site only for brief bouts each spring and fall, when the sand was moist and the wind was generally calm. The excavation continued intermittently from 1968 to 1977. The burial features yielded information about an extensive array of utilitarian, decorative and food-related items that had been placed with the dead. There were other, similar, cemetery sites in the region, but most were discovered as they were destroyed in the process of sand and gravel extraction, deep plowing, or other landscape modifications. Collectively, they have been described as evidence of the "Glacial Kame Burial Complex" (Cunningham, 1948). This is an archeologically defined cultural group within the Late Archaic period. As noted in Chapter 3, these communities used a broad range of tools and competencies, including control of fire, to successfully live off the land. Their subsistence strategies predate reliance on domesticated plant foods and the manufacture of pottery. Late Archaic groups moved around the landscape, probably in seasonal patterns, exploiting naturally occurring foodstuffs.

So little to go on

What do we know about these people who made their home in the untamed, postglacial landscape? Not much. We know that they lived in a well-watered deciduous forest, with a stable, temperate climate (Karrow & Warner, 1990). The range of foods would have been extensive, although seasonal in availability (Monckton, 2013; Needs-Howarth, 2013). Small bands are thought to have moved seasonally, following routes determined by resource availability (Ellis, Timmins, & Martelle, 2009). There is a kind of working hypothesis that springtime aggregations at prime fishing locales would have been part of the annual cycle (Wright, 1963) but evidence is ephemeral. Perhaps the absence of evidence is because people lived along stream edges and marshes (Woodley, 2006). So far, two sites provide some evidence of housing. The Davidson Site is located about 100 km (62 mi) from the Hind Site and dates from about 3200–2800 years ago. There, house floors were dug below the natural surface, creating semi-subterranean shelters. The effort made to take advantage of the insulative value of the earth suggests that this kind of shelter may have been used in winter (Ellis, Keron, Menzies, Monckton, & Stewart, 2015).

In addition to burial gifts found in cemeteries, we can learn about the people from the distinctive stone projectiles and tools that are sometimes found on the surface of the land. The Glacial Kame people transformed stone into implements by flaking and by grinding. We know they created and used projectile points, net sinkers, bannerstones (probably used as weights on the handles of spears), tools for woodworking, and they modified animal bones and teeth into tools for leatherwork, food preparation and adornment. The designs of many of the stone pieces show impressive balance, symmetry and creative use of the available raw materials. Subjectively speaking, much of what we can see of the material culture is beautiful. We know that people traded raw materials over great distances. Small items for personal adornment were made from hammered copper. Copper items were included with some burials, the ore having been traded down from the northern margin of Lake Superior (Donaldson & Wortner, 1995).

Using ethnographic analogy with historic groups elsewhere plus this kind of fragmentary evidence from sites and artifacts, a picture forms of Glacial Kame family groups moving regularly, setting up camps in different locations depending on when food or other resources would be available. Archeologists call this residential mobility. Alternately, they may have had more permanent home bases, with subsets of the group (individuals or pairs) ranging out to hunt, fish, trap, collect and so on. This is called logistic mobility. Logistic mobility is seen as more efficient than residential mobility, in which everyone moves each time. Much of this is just surmise. At present, we do not know much about their movement patterns. For the story I want to tell you, it is important to keep in mind what we know these folks did not have. There is no evidence that they stored significant amounts of surplus food. They did not exploit domesticated plants or animals (except perhaps dogs). They did not regularly make pottery and they did not use wheels. This might have been a rich and bountiful life for healthy, vigorous people. It would have been very difficult indeed if you could not walk.

The Hind Site

The archeological features excavated at the Hind Site include 22 primary burials (in which the skeleton is complete, and the bones are articulated as in life) and ten cremation features. The cremations were not in-place incinerations, but rather were roundish parcels placed in holes, sometimes with an artifact placed on the bundle, always with red ochre (ground hematite) added, then covered. The condition of the calcined bone and tooth fragments shows they were burned when the bone was fresh. Fresh bone is both wet and greasy, so when exposed to extreme heat it shrinks, twists, and breaks at angles unlike what is seen when dry bone is burnt. Fresh teeth explode when exposed to high heat, as the moisture in the pulp cavity expands (Mayne Correia, 1997). When examined for small identifiable bits, those ten burials of cremated remains represent at least nineteen people. The presence of fragments from unerupted and developing tooth crowns indicate the presence of infant and child remains. The extensive wear on some teeth indicates the presence of relatively elderly persons. So, cremation was used for all ages, and sometimes the remains of more than one person were combined in a bundle.

The 22 primary burials, like the cremations, commonly included generous amounts of red ochre. This category included two infants who had died during the first year of life, two people who died in mid-to-late childhood, and two in their mid- to-late teens. The sixteen adults included seven women, seven men and two individuals represented by remains that were too fragmentary for us to confidently attribute sex. The adult ages at death range from the early adult decades to more than 60 years. Taken together, it looks like this was the burial site for all group members who died, with cremation used when those deaths took place at a distance (Pfeiffer, 1974). We don't know how long the cemetery was in use, so the sample does not help us reconstruct group size. We also don't know how closely related they were, although the presence of a rare anatomical anomaly in three adults suggests family ties. In those three, the cartilage around the thyroid gland in the neck had ossified. That is, the cartilage had been replaced by bone. It would have had no significance to their health, but it may indicate a shared genome.

The preservation of most skeletons was exceptionally good, thanks to the well-drained, very sandy matrix with very little plant root activity, and, in some cases, the copper artifacts. The salts formed by weathering copper are both a desiccant and a toxin. When copper is in contact with a body, it leads to green staining of bone and sometimes small areas of dried skeletal muscle and skin.

The bones and teeth provided information about the people buried there. Ages at death for the children were determined from dental formation. Based on these ages, most of the children show linear growth that is within the normal range for healthy children. Radiographs (x-rays) of their long bones show growth arrest lines in three of the six children. These lines of opacity, perpendicular to the shaft's length, are thought to reflect periods of ill health or significant malnutrition, followed by recovery. The bone-growth cells essentially "march in place" for a bit, until the body recovers enough to resume growth. It is reasonable to conclude that most of the children died of acute circumstances, like accidents, exposure, animal attacks and the like. The adult skeletons have prominent muscle attachment sites and considerable osteoarthritis to their joints—especially to their vertebrae. These features are consistent with physically vigorous and challenging lives. Among these skeletons, there is no evidence of healed fractures or bone damage from interpersonal violence. As with most children, causes for the adult deaths remain unknown.

The teeth and jaws of the Hind Site skeletons are especially distinctive. The chewing surfaces of the permanent teeth show a lot of wear. The cusps of the back teeth were worn flat by the time someone reached early adulthood. The older adults who still had teeth when they died show dentitions that have simply worn out. When teeth are worn too quickly for the formation of secondary dentine to keep up, the pulp cavity becomes exposed. That can lead to infection and loss of the tooth through abscessing. This was often the case here. Sometimes there is no sign of an abscess, but the tooth material has simply ground away. On surviving teeth, there are chips along the edges and excess cementum surrounding the root, the latter caused by persistent vigorous use. Thanks to meticulous excavation, we could see that sometimes the bone along the gum line had resorbed to obliterate the tooth socket, yet small bits of tooth root remained within the mouth. Those bits would have been held in place by soft tissue, perhaps still providing some value for processing food. What could have required such vigorous mastication? Gifts buried with the deceased include the bones of mammals (deer, bear, raccoon and others), reptiles and birds, freshwater mollusks, nuts and berries. Food options would have included both lacustrine and riverine fish, leafy plants like sumac and cattail, fleshy fruits like plum, cherry, blueberry and others, and a diversity of nuts including butternut, walnut and hickory. While one cannot know with confidence, I think that transportable, daily staples could have been dried meat and fish as well as nuts.

A child's distinctive skeleton

Within this assemblage, one skeleton is particularly distinctive. Rather than being robustly built, it is markedly, pathologically delicate. It is the skeleton of a child buried on her side, tightly nestled against an adult woman (Vairamuthu & Pfeiffer, 2018, Fig. 2). The child was buried in a single burial event with two other group members. The first item that was placed into the burial pit was a bag, now disintegrated, that held fragments of cremated bone mingled with red ochre. Later examination suggests that these were the remains of one young adult, perhaps a woman based on the relative size of the identifiable fragments. Perhaps the inclusion of red ochre within the receptacle indicates that a burial ritual took place soon after the body was cremated at a locale some distance from the cemetery. The fragments may have been carried by a group member for many months.

Immediately above this were two primary burials in a shared a grave—an older woman and the young child. The head of the child was placed over the packet of cremated fragments. The two complete skeletons were placed on their sides, facing one another, with their limbs tightly flexed. Their closely clustered bodies formed an oval. Their feet were placed at the base of the grave, which meant that the child's head rested against the woman's knees. After their placement, red ochre had been sprinkled on both bodies, with concentrations on the facial areas, as well as the lower torso and ribs of the child. The woman was wearing a bracelet of more than 60 small copper beads on her right wrist. Beneath her hand, the copper salts and the ochre helped to preserve an oval-shaped packet of foodstuffs: several hickory nuts and the bones of an unidentified reptile. Their respective ages make it plausible that they were mother and child.

Study of skeletal indicators of age at death suggest that the woman had lived for at least forty years, perhaps considerably more. She had osteo-arthritic changes to her wrists, elbows, shoulders, knees and particularly notable arthritic lipping to her spine. During life, she had lost all the teeth from her lower jaw, and all the back teeth from her upper jaw. The almost complete loss of the woman's dentition probably reflects a much older age at death than our skeletal assessment methods indicate. A radiocarbon date was obtained from a nearby burial with similar copper beads (BP 2875 ± 75; S1061 (Wilmeth, 1978)). That measurement must be calibrated to account for recent anthropogenic wobbles in the ratio of carbon-14 in our environment, like the burning of fossil fuels, as well as nonanthropogenic wobbles. Using current calibration (CALIB REV7.1.0 (Stuiver & Reimer, 1993)) the median probable date is 1060 BCE. That is about 3000 years ago.

It was obvious immediately that the child's skeleton was unusual. Her bones were extremely delicate, and her mouth had few teeth and little bone to support them. The child is not as young as her size initially suggested. The developmental stage of her few remaining teeth is consistent with an age of 14.5–16.5 years (AlQahtani, Hector, & Liversidge, 2010). Her second permanent molars (which erupt at about age 12) were fully formed and her third molars (wisdom teeth) were poised to erupt (Vairamuthu & Pfeiffer, 2018, Fig. 4). The developmental stage of the skeleton is consistent with what would be expected of a person aged about 16 years. It is based on this teen-aged dental and skeletal maturity that I use a female pronoun. The skeletons of young children all tend to look feminine, but after puberty begins, there is differentiation, especially of the face and hips. This skeleton retains a feminine shape.

The girl's skeleton shows both stunting and wasting. No portion of the skeleton was spared from the condition. The length of the femur, for example, is parallel to expectations for a 6–8-year-old child. The diameters of the long bones are about 40–50% of adult values, at an age when they should be much nearer to adult size. Applying estimates developed from normal skeletons (Visser, 1998), she would have been about 120 cm (47″) tall, weighing about 22 kg (50 lb).

The delicacy of build is pervasive and appears to reflect both a shortage of bone formation and a loss of bone (Box 4.1). The bone loss is illustrated especially by the jaws. Most of her teeth were lost in life, because of the loss of the surrounding bone. On both sides of the mandible where the first

BOX 4.1 Bone tissue: Modeling and remodeling

The archeological discovery of a skeleton is possible because skeletons are made of bone tissue, which has a mineral component. Bone is a remarkable material that is both strong and somewhat flexible. It is shaped in ways that are custom designed to provide support for a human-sized body that performs human-scale activities. It also provides reliable sites for soft tissue attachments, protection for crucial organs and activities, and an emergency supply of calcium and phosphorous. In life, bone is 20% water, 30–40% organic and 40–50% inorganic material. The organic part is mostly (90%) collagen—a protein. The inorganic part is a form of calcium phosphate called hydroxyapatite: $Ca_{10}(PO_4)_6(OH)_2$. From birth to death, a person's bone tissue will vary in its stiffness, brittleness and mass, but bone is bone. The tissue components are always the same.

Bone is initially formed when mineral pervades a collagen template, incorporating nerves and blood vessels within the developing framework. The creation of new bone is called modeling. Even as the first modeling occurs, remodeling needs to begin. Think of the growing end of a long bone as bugle shaped. The wide bell needs to extend as the bone gets longer, but the prior width of the bell needs to be narrowed, as the former bone end becomes the shaft. Some bone tissue must be removed, in synchrony with the creation of new tissue. That is called remodeling. The tempo and coordination of bone modeling and remodeling are sensitive to environmental factors. Serious health challenges in childhood can cause bone growth to cease, with potential catch-up when normal conditions return.

Both modeling and remodeling are executed by two cell types: osteoblasts and osteoclasts, forming and removing bone. To the frustration of learners who would like "blasts" to destroy because it would be an easy mnemonic, the "blasts" actually do the building. The "clasts" do the destroying. Once maturation is achieved (as late as ages in the mid-20s for some areas of the skeleton), modeling mostly ceases—unless needed to heal a break—but remodeling continues throughout adulthood. Bone is living tissue, and there needs to be maintenance remodeling throughout life. There are mechanical causes, like building additional bone in response to exercise or losing bone because of disuse. There can also be hormonal causes, like postmenopausal osteoporosis.

BOX 4.1 Bone tissue: Modeling and remodeling—cont'd

There are many diseases and traumatic events that can affect the skeleton. Their investigation, termed palaeopathology, is a fascinating field. It is also a challenging one, thanks to bone's limited types of response. Bone can deposit and resorb. That's it. The potential cause for abnormal bone shape, size or texture must be ascertained through consideration of the evidence. It requires an understanding of bone modeling and remodeling, but also knowledge of soft tissue anatomy. In many instances, the bone has remodeled in response to soft tissue that has long since disappeared.

molars would have been, the body of the mandible is only about 1.5 cm (less than an inch) in height. Also notable is the almost complete lack of tooth wear on the few remaining front teeth and the second molars. Some of the teeth show substantial buildup of calculus (plaque).

There is evidence from both the upper and lower limbs that this young girl did not move normally. Bone strength of the upper limbs, as determined from cross-sectional geometry, is less than half of that expected for a child of her age. Perhaps most intriguing, the research of Thivviya Vairamuthu demonstrated that the bicondylar angle of the femur (at the knee) is consistent with that of infants (Vairamuthu & Pfeiffer, 2018). A distinctive new alignment of the bones that meet at the knee develops when a child begins to walk. Standing and walking load the lower limbs in a manner, and with an intensity, that stimulates new alignment (Fig. 4.2). Based on the skeletal evidence, the child did not walk and did not regularly use her hands and arms for vigorous tasks. Her wedge-shaped lower vertebrae (Vairamuthu & Pfeiffer, 2018, Fig. 3) suggest that she may have been limited to a seated, crouched posture.

What medical conditions might lead to this constrained bone growth, leading to extreme delicacy of bone, and loss of bone around the tooth roots? In the long bones, the ratio of the cortical (dense) to trabecular (spongy) bone is as expected, but the total amount of bone is less than expected. This indicates osteoporosis, which is a symptom, not a cause. Vairamuthu and I explored the differential diagnosis. Smooth and slender long bone shafts and low bone mass are associated with paralysis and other causes of disuse atrophy, but in this case the entire skeleton was affected, including the skull and jaws. There are no nutritional deficiencies that cause the pattern seen here, nor does the pattern follow what is expected in many hereditary disorders.

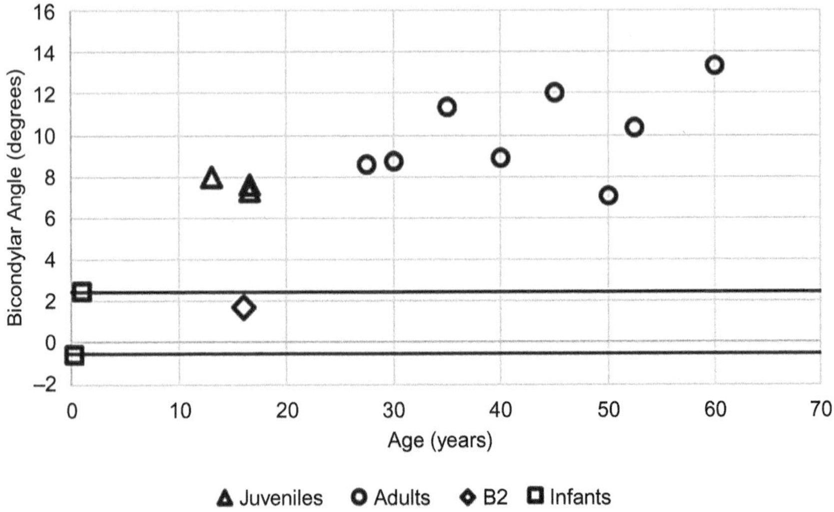

Fig. 4.2 A graph showing the angle of the femur at the knee in infants (bounded by heavy, horizontal lines), children and adults. The delicately built child (labeled B2) has an angle like that of an infant, suggesting that her legs had not been used for normal locomotion. (Credit: T. Vairamuthu.)

We concluded that a form of osteogenesis imperfecta (OI) was the most probable explanation. This is an inherited disorder that is quite variable in its expression. Some types of OI are associated with multiple fractures and deformed bones, but others are not. Type IV is most consistent with this child's skeletal features (Vairamuthu & Pfeiffer, 2018).

Information from modern cases of OI suggests that she may have suffered several nonskeletal symptoms, including hearing loss and she may have bruised easily. Modern studies also note that people with OI have a particular behavioral feature of notable intelligence. They tend to be bright, with resilient temperaments (Ablon, 2003). From infancy onward, this child would have required help with self-care as well as diet and mobility. The condition of her teeth and jaws indicates that her diet was much softer than that of other members of the group. Assistance would have been needed for diet preparation, transportation, hygiene, and various activities requiring arm strength. Both meat and nuts can be ground and mixed with water, but this and other supportive tasks would have required significant commitment by one or more caregivers. As the group moved from one location to another, she would have been carried by someone. She would not have been able to participate in food gathering and other subsistence activities. On the other hand, based on parallels with modern cases, she may have made valued intellectual contributions to her community. We cannot say what led to her death, nor do we know the nature of her relationship with the two women with whom she was buried, one of whom died at about the same time. What we can say is that she lived for many years. Her survival reflects the vigilant, creative, sustained efforts of the people around her.

Looking back at the Hind Site

From the outset, the careful excavation and excellent preservation of the Hind Site skeletons attracted interest from archeological scientists. In the early days, the information available from different scientific approaches was rather unpredictable. Four conventional radiocarbon dates (Box 1.2) were run in the 1970s (Wilmeth, 1978). In those days, each date required a large sample of human bone—typically one half of an adult rib cage. In those days, it was assumed that the ribs had no diagnostic value regarding a person's habits and health, so this destruction was seen as justified. One rib was sent to researchers exploring the new idea of using stable isotopes for dietary information, but that lab failed to extract sufficient collagen. A small fragment of rib was sent to another scientist for early research into ancient DNA extraction. Many years later, after years of improvements in lab methods, the collagen from that rib yielded a fragment of mitochondrial DNA from that person. This helped to begin the exploration of relationships between ancient Indigenous North Americans and living Indigenous groups (Shook & Smith, 2008). Today, a single tooth would be sufficient to provide a date, as well as dietary isotopes and aDNA information. Five teeth from Hind Site skeletons have recently been included in an ancient DNA initiative mounted by a consortium in Canada and the United Kingdom, the results of which are not yet available.

The Hind Site has been a part of my life for almost 50 years now. The bones are curated with great care at the University of Toronto, Department of Anthropology. As a part of her research project, Vairamuthu sculpted high density foam for each bone of Burial 2, to give them extra protection. Radiographs are also on file. The Walpole Island First Nation is the Indigenous group that lives closest to Bothwell, Ontario. They are aware of the collection and they are informed about the results of all new studies. I think they know how important it is to learn more about these early occupants of their region. In 2012, Stanley Wortner was awarded a medal by the Ontario Archeological Society for his significant contributions as an "avocational archaeologist extraordinaire" since 1948.

I hope that my description of the history of the collection, from discovery to the present day, shows that "studying the bones" is an activity that changes in focus and direction as scientific methods improve. We are able to ask questions about past people now that we could not hope to answer in past decades. For example, from the time I first encountered her, I wondered about the child known as Burial 2. Even the first description—in my PhD thesis—described the skeleton as unusual and certainly pathological (Pfeiffer, 1977). Attempts to give a name to the pattern failed until we had access to an expert on-line community, as well as information retrieval systems that have been a boon to palaeopathologists. Details of our analysis were dependent on relatively new knowledge about the normal development of children's skeletons. In recent years, palaeopathologists have broadened and deepened their scope well beyond differential diagnosis. The field now gives serious attention to the "bioarchaeology of care", where effort is directed to determining not only the diagnosis, but also the level and type of disability and the model of care that would have been needed to sustain the person's life (Tilley, 2015; Tilley & Cameron, 2014).

The existence of this collection and others like it form the basis for ancient DNA studies about the peopling of the Americas. In many parts of North America, the evidence of the people who lived off the land prior to the transition to agriculture is very sparse. That evidence is therefore particularly precious. Just as important, evidence from collections like the Hind Site provides a window into the lives of past people. The facts challenge the Hobbesian assertion that the life of "natural man" was "solitary, poor, nasty, brutish and short". Modern life provides little to no experience with making a tool kit from stone and bone, hunting, setting trap lines, monitoring plant maturation, walking long distances with all of one's possessions in tow, year after year. Yet past people, just like us, secured food and shelter, undertook large social gatherings, followed important traditions—like the rituals of burying the dead. As illustrated by the skeleton of a child who survived for many years despite a significant disability, the hunter-gatherers of the distant past could also stretch their capacities to care for family members with special needs. It happens today, and it happened 3000 years ago.

References

Ablon, J. (2003). Personality and stereotype in osteogenesis imperfecta: Behavioral phenotype or response to life's hard challenges? *American Journal of Medical Genetics. Part A, 122A*, 201–214.

AlQahtani, S. J., Hector, M. P., & Liversidge, H. M. (2010). Brief communication: The London atlas of human tooth development and eruption. *American Journal of Physical Anthropology, 142*, 481–490.

Cunningham, W. M. (1948). *A study of the glacial kame culture in Michigan, Ohio and Indiana. Vol. 12.* Ann Arbor: University of Michigan.

Donaldson, W. S., & Wortner, S. (1995). The Hind site and the glacial kame burial complex in Ontario. *Ontario Archaeology, 59*, 5–95.

Ellis, C. (2013). Before pottery: Paleoindian and Archaic hunter-gatherers. In M. K. Munson, & S. M. Jamieson (Eds.), *Before Ontario: The archaeology of a province* (pp. 32–47). Montreal & Kingston: McGill-Queen's University Press.

Ellis, C., Timmins, P. A., & Martelle, H. (2009). At the crossroads and periphery: The archaic archeological record of Southern Ontario. In T. E. Emerson, D. L. McElrath, & A. C. Fortier (Eds.), *Archaic societies: Diversity and complexity across the midcontinent* (pp. 787–824). Albany, N.Y: State University of New York Press.

Ellis, C. J., Keron, J. R., Menzies, J., Monckton, S. G., & Stewart, A. (2015). For immediate occupancy: Cozy three-thousand-year-old heritage winter house with river view near Lake Huron—apply to Terminal Archaic realty. In B. G. Redmond, & R. A. Genheimer (Eds.), *Building the past: Prehistoric wooden post architecture in the Ohio Valley—Great Lakes* (pp. 29–62). Gainsville: University Press of Florida.

Karrow, P. F., & Warner, B. (1990). The geological and biological environment for human occupation in Southern Ontario. In C. Ellis, & N. Ferris (Eds.), *The Archaeology of Southern Ontario to A.D. 1650* (pp. 5–35). London, Ontario: London Chapter, Ontario Archaeological Society.

Mayne Correia, P. M. (1997). Fire modification of bone: A review of the literature. In A. H. Haglund, & M. Sorg (Eds.), *Forensic taphonomy: The postmortem fate of human remains* (pp. 275–294). Boca Raton: CRC Press.

Monckton, S. G. (2013). Plants and the archaeology of the invisible. In M. K. Munson, & S. M. Jamieson (Eds.), *Before Ontario: The archaeology of a province* (pp. 124–133). Montreal & Kingston: McGill-Queen's University Press.

Needs-Howarth, S. (2013). Animals and archaeologists. In M. K. Munson, & S. M. Jamieson (Eds.), *Before Ontario: The archaeology of a province* (pp. 112–123). Montreal & Kingston: McGill-Queen's University Press.

Pfeiffer, S. (1974). Analysis of cremations from two Late Archaic burial sites: Riverside and Hind. *Wisconsin Archaeologist, 55*(4), 302–311.

Pfeiffer, S. (1977). *The skeletal biology of Archaic populations of the Great Lakes region. Vol. 64.* Ottawa: National Museums of Canada.

Shook, B. A. S., & Smith, D. G. (2008). Using ancient mtDNA to reconstruct the population history of northeastern North America. *American Journal of Physical Anthropology, 137*(1), 14–29.

Storck, P. L. (2004). *Journey to the ice age: Discovering an ancient world.* Vancouver, B.C.: UBC Press.

Stuiver, M., & Reimer, P. J. (1993). Extended C14 data base and revised calib 3.0 C14 age calibration program. *Radiocarbon, 35*, 215–230.

Tilley, L. (2015). *Theory and practice in the bioarchaeology of care.* Springer.

Tilley, L., & Cameron, T. (2014). Introducing the index of care: A web-based application supporting archaeological research into health-related care. *International Journal of Palaeopathology, 6*, 5–9.

Vairamuthu, T., & Pfeiffer, S. (2018). A juvenile with compromised osteogenesis provides insights into past hunter-gatherer lives. *International Journal of Palaeopathology, 20*, 1–9. https://doi.org/10.1016/j.ijpp.2017.11.002.

Visser, E. P. (1998). Little waifs: Estimating child body size from historic skeletal material. *International Journal of Osteoarchaeology, 8*, 413–423.

Wilmeth, R. (1978). *Canadian archaeological radiocarbon dates. Vol.* 77. Ottawa, Ontario: National Museum of Canada.
Woodley, P. J. (2006). Archaic occupations of the Peiganovitch Site, South-Central Ontario. *Ontario Archaeology, 81*(82), 44–72.
Wright, J.V. (1963). Archaeology of the Donaldson Site. *National Museum of Canada Bulletin, 184*.VII-91.

CHAPTER 5
We enter the afterlife together

Prologue

When someone dies, people close to that person do their best to follow their wishes for treating the body. While the specific practices can be quite diverse, mortuary traditions convey respect and continuity between life and death. They show impressive continuity through time. It is not surprising that anthropologists have given the topic a lot of attention.

Seventeenth century missionaries from France were particularly attentive to the ideas and actions of their local hosts, the Huron-Wendat Nation, with respect to death and commemoration. The detailed reports of Jesuit missionaries and French explorers are particularly thorough, and relatively unbiased, as these things go. Some of the Jesuits stayed with communities for years, gaining language mastery and observing events that happened only rarely. Their letters from New France, 1632–1673 CE, are known as the Jesuit Relations (see Chapter 3).

In May 1636, Jean de Brébeuf penned a first-person account of the Feast of the Dead at a village called Ossossané. This was a multiday celebration of those who had died during the prior decade or so. The Jesuit missionary had spent several years with the community, so he knew them and their language quite well. His account was both thorough and culturally sympathetic (Seeman, 2011:59). With friendly neighbors, the grieving community constructed a single, deep pit, lined it with deer skins, surrounded it with scaffolding to display their dead. The reverent handling of the deceased may have seemed somewhat akin to Brébeuf's experience with Christian relics back in France. Men positioned in the pit indicated the timing and placement of the remains. Items of value were placed in the pit, to accompany the souls on their journey. Because the dead had previously been interred elsewhere, we call this a secondary burial. The pit itself is called an ossuary. While the term "ossuary" commonly means a container or room in which bones of the dead are placed, here the term refers to a mode of mortuary practice. Gifts were placed within the pit, and priests standing in the pit purposely co-mingled the deceased. Their souls were thus sent on their way,

Osteobiographies
https://doi.org/10.1016/B978-0-12-823880-6.00006-9

on a collective journey to their next home. The living, having paid their respects, could re-locate their village if needed, to somewhere with fresh, fertile ground for their crops, or better fishing and hunting.

Background to the story

To get an idea of the social context within which a person lived, it is helpful to establish some sense of the broader context, through space and time (Fig. 5.1). The story of the origin and historic development of the Huron-Wendat Nation is a synthesis of research by archeologists, linguists, historians and paleogenome scientists. The hunter-gatherers like those described in Chapter 4 appear to have been ancestral Algonquian-speakers. Cultural continuities, evidence from ancient mtDNA (Shook & Smith, 2008) and language analyses support this view. The people who settled into larger communities and relied on farming spoke a different, unrelated language: Northern Iroquoian. The two languages do not have much in common, linguistically speaking. This led scholars to hypothesize that a population

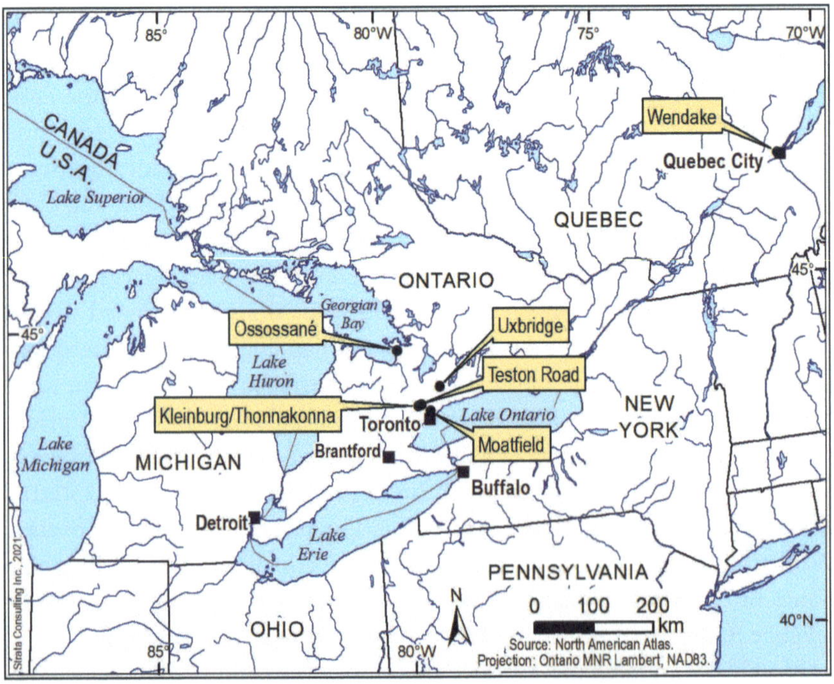

Fig. 5.1 A map of the region discussed in this chapter.

of maize farmers (or their predecessors) moved into the region from elsewhere. This is known as the migration hypothesis, or, in the parlance of paleogenomics, demic diffusion. The problem with the migration hypothesis is that there is no evidence of inter-group conflict at the time it would be expected, and studies of stable isotopes linked to diet show that the adoption of maize as an important food is very gradual.

Current thinking supports a different perspective: the in-situ hypothesis, or cultural diffusion. This imagines that a subset of ancestral Algonquian speakers adopted both farming and the language of farming, from Iroquoian-speakers further south. Growers would have had to select for maize that could mature in a shorter growing season, in order to successfully farm in more northerly fields. Seeds adapted to southerly climes would not mature fast enough to beat the autumn frosts. Obviously, there would have been contact between the groups and some co-mingling, but not wholesale population replacement. The palaeogenomic information available so far, based on ancient mitochondrial DNA (Pfeiffer, Willamson, Sealy, Smith, & Snow, 2014), is consistent with the in-situ story line.

The Northern Iroquoian farmers had Algonquian speaking neighbors to the east, west, south and north. Archeological evidence shows that Iroquoian communities were in conflict with one another, and with neighboring Algonquian nations, from the thirteenth century onward. By the historic period, the Wendat and their Algonquian neighbors were allied with the French and were at war with the Haudenosaunee, a war that intensified in the mid-sixteenth century. The alliance with their Algonquian allies was especially important to the Wendat, as they traded maize in exchange for deer skins and beaver pelts, which were not available in sufficient quantities within their home ranges (Birch & Williamson, 2013).

The full Northern Iroquoian cultural complex, like that described in the early European reports, is not archeologically recognized until the turn of the fourteenth century. The prior centuries saw gradually expanding populations of farmers who also relied on hunting and fishing (Warrick, 2008). Earlier communities lived close to the northern shore of Lake Ontario, but people later moved further north to the southern shore of Georgian Bay—a major bay of Lake Huron. Aspects of distinctly Huron-Wendat culture arose at different times. Purposely comingled ossuary burial can be dated back to the thirteenth century. It is interesting that some other Iroquoian speaking farmers to the south also interred their dead collectively. However, in those cases individual bodies were placed next to one another rather than being comingled (Williamson & Steiss, 2003). The Feast of the Dead and the creation of a truly collective grave, as created and regularly enacted by the Huron-Wendat, was a unique cultural feature.

The excavation of ossuaries

As the province of Ontario was developed for farming by immigrant newcomers—most from the British Isles—the existence of ossuaries on the landscape was a source of excitement and wonderment. In the centuries since their creation, the soft tissue would have decayed away, so each ossuary would have become a more or less round, shallow depression on the landscape. The pits' contents were often explored after a farmer's plow exposed dry, weathered bone. "Race science" was a dominant intellectual pursuit in the nineteenth century, so digging ensued, with the goal of finding "racially diagnostic" crania. Two commonly held assumptions helped to fuel the destruction of ossuaries: the presumed scientific value of the well-preserved skull, and the presumed presence of curios that were thought to be comingled with the bones. In 1937, Sir F.S.H. Knowles was able to report measurements from over 150 adult crania taken from ossuaries discovered during the prior 50 years (Knowles, 1937). True to that era, his study gave no attention to juveniles and scant attention to bones from below the neck, except to establish adult body size from leg long bone lengths. Like other scholars before him, he reported "a confused mixture of cranial types" (1937:53).

The ossuary described by Brebéuf—Ossossané—was excavated in 1947. Archaeology had been established as a taught discipline at around that time. The mode of ossuary exploration was one that compromised the bones and prioritized any- and everything else. The ossuaries created by the ancestral Wendat people were about two meters deep, but excavators followed hearsay of other diggers that the skulls would be on top. This, and technical limitations, meant that the excavation mode was one of sitting atop of the ossuary and digging down. Professional archeologists did this in a controlled manner, identifying arbitrary layers and leaving unexcavated walls between units in order to see the profile of the ossuary. Nevertheless, the effect of one or more living bodies pressing down on several feet of comingled bones with relatively little soil packed around them for support was inevitable, extensive, breakage of those bones. It was not until the excavation of the Moatfield ossuary (located in north-central Toronto) in 1997 that an ancestral ossuary was excavated as a pedestaled unit, with excavators' weight outside the deposit of bone (Williamson, Thomas, & MacDonald, 2003).

This digression into excavation methodology may seem like a petty concern, but it is not. Ossuaries were built to hold the comingled remains of a large number of people—from dozens to hundreds. Ossossané held a minimum of 419, based on bone counts after excavation (Pfeiffer & Fairgrieve, 1994). Each skeleton has a minimum of 206 bones and associated teeth. Even if the bones remained whole and in proximity to their mates, it would be very difficult to reconstruct individual skeletons. Once they were broken by pressure from above—first the plow, then the archeologist—and they were removed from the ground in arbitrary spatial units, they became something quite different from a group of skeletons. They became a collection of co-mingled, broken bones. This is why, despite the professional excavation of over twenty ancestral Huron-Wendat ossuaries (Williamson, 2014), few have yielded much bioarchaeological information.

Ossuary osteobiographies

The first attempt to create something like an osteobiography from the bones removed from an ancestral Huron-Wendat ossuary was published in 1963. James Anderson's monograph, *The People of Fairty* (Anderson, 1963), was highly regarded as a new approach to the study of past peoples. Its very name suggested that there was more to learn than just skull type. In those days, radiocarbon dating was rarely done. The site was excavated by professional archeologists (J.N. Emerson and J.V. Wright) in 1956. There had been no artifacts found, so there was no sense of how long ago the People of Fairty (minimally 512 of them) had been interred. We now know that Fairty dates to 1365–1385 CE, so its construction predates Ossossané by almost three centuries.

Anderson sought to describe a community, as best he could. In addition to the usual measurements of adult crania and long bones, he focused attention on anatomical variants that may carry genetic information, forming a basis for comparison with other skeletal collections through which population relationships might be reconstructed. The migration hypothesis was in play, so there was great interest in where the farmers had come from. Anatomical evidence for health problems like premortem tooth loss, healed bone fractures and osteoarthritis were enumerated as well. In this category, he described a few bones—mostly vertebrae—that showed "cavitation with abscess formation" of the kind that is sometimes caused by *Mycobacterium tuberculosis*, although he did not presume to offer that diagnosis. In those days, it was believed that tuberculosis had been introduced to

North American peoples by immigrating Europeans. After all, tuberculosis had caused epidemics among Indigenous Americans from the seventeenth century onward. Hence, what Anderson and others had seen among bones from ossuaries could not reflect tuberculosis, they believed. Over the next two decades, that perspective would change.

A central feature of Anderson's work was the pie chart mortality profile (Fig. 5.2). Anderson sought to categorize the age-at-death of every person buried at Fairty, presumably from dental development and other indicators, although his methods are not described. In that era, the construction of life tables like those used by demographers was gaining popularity.

Demography is the study of different life-history characteristics of a population, such as size, kinship and migration. Palaeodemography, as it was known, was pursued because it might demonstrate an impact on life expectancy of factors like major changes in diet or the arrival of Europeans and their diseases. By the mid-1980s, life tables had been constructed for eight ossuaries (Jackes, 1986). They consistently showed low life expectancy, with virtually no one living beyond 50 years. Were there no grandparents? Recognition of the osteological paradox (see Chapter 1) and the shortcomings of estimating older adults' ages-at-death from single, isolated bones dampened enthusiasm for this research approach (Bocquet-Appel & Masset, 1982). Palaeodemography was not likely to fulfill advocates' hopes. Without accurate adult age-at-death estimates, there could be no exploration of how

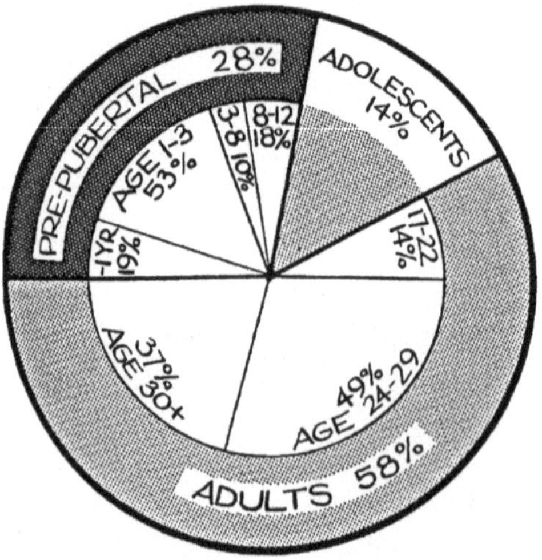

Fig. 5.2 Age distribution of 512 individuals from the Fairty Site ossuary (Anderson, 1963:33).

historic events affected adult life expectancy. Nevertheless, giving attention to all skeletal remains, regardless of age-at-death, was a positive development. It was a first step toward studies of infants' diets, mothers' decisions about weaning, and children's growth and development. These aspects of life and health were all things that also might reflect changing community circumstances through time.

One consistent paleodemographic feature seen in all ancestral Huron-Wendat ossuaries is a shortage of very young infants. In any community living without access to modern medical interventions, it is expected that some infants will fail to survive the first year of life, even if there is attentive care and sufficient diet (Bourgeois-Pichat, 1951; Lewis, 2007). The ossuaries included proportionally fewer newborns than would be expected. This is consistent with what the Jesuits observed. Ethnographer Elizabeth Tooker described evidence from the Jesuit Relations (JR) like this: "If a young baby, less than a month or two old, died, it was buried by the road, in order that it might enter a woman's womb and be reborn (JR 10: 273; cf. JR 15:183—"They believe that souls enter bodies after death.")" The Wendat ossuaries of several centuries demonstrate this practice, and the frequent discovery by archeologists of newborn remains under Wendat house floors reinforces the observation. The practice reflects the high value communities placed on children and fertility. It is a reminder that the decisions about every person who passed away were based not only on personal feelings of grief, but also on the traditions of the community.

Evidence of a disease that wasn't supposed to be there

Sometimes the presence of extensive written records can throw you off. To the extent that the Jesuits wrote about public health, they said the Huron-Wendat were tall, robust, and "more healthy than we" (Thwaites, 1896, Vol. 38:257). Neither chronic food shortages nor common diseases were noted. What, then, does the bioarchaeologist make of abnormal bones, the size and shape of which were altered by prolonged pressure from soft tissue that was present where it shouldn't have been? A small number of bone surfaces at locations of joints and intervertebral disks were modified in ways that were not caused by postmortem damage. Some of the vertebrae had collapsed because of bone loss and had fused together with the vertebrae above and below, creating a sharp forward angle to the spine (kyphosis). The mid-back location and structure of the fused bones of the spine would have caused what is known as a "dowager's hump." These days, this is sometimes seen as the result of vertebral fracture from osteoporosis. In the archeological bones, the bone tissue beyond the focus of damage is not osteoporotic. The question of what caused such remodeling vexed researchers studying Ontario ossuaries, as well as other scholars studying the bones of Indigenous pre-European-contact farmers from other parts of the Americas.

The fundamental method of palaeopathology is differential diagnosis. Abnormal features are described without presumption of cause, then conditions that might cause such features are identified. From there, a contextual, probabilistic exercise seeks likely causal candidates. Researchers applying differential diagnosis to the modified bones of spines, hips and occasionally other joints kept landing on *Mycobacterium tuberculosis* (TB) as the most likely causative agent. But surely the Jesuits would have known TB, which was well known in France at the time, if they encountered it. Surely, they would have mentioned it in their letters.

In a fine example of how science works, evidence accumulated, and scholars sought to make sense of it. In 1981, a small edited volume hoisted a flag in favor of acknowledging the presence of prehistoric tuberculosis in the Americas (Buikstra, 1981). One of the chapters described evidence from an Ontario ossuary. The ossuaries of the Northern Iroquoians provided many examples. Still, their comingled nature made interpretation difficult. Tuberculosis of the familiar European variety is known to affect bone tissue in fewer than 5% of cases, most commonly along the spine as the infection spreads from the lungs. Bone involvement occurs only after the pathogen is well established within the body. It seemed like there were too many affected vertebrae, too many people affected. In 1984, I applied epidemiological methods to describing affected bones from the Uxbridge ossuary (Pfeiffer, 1984). This fifteenth century burial site located just north of Toronto had held at least 457 people. It included at least 26 skeletons affected by what looked like tuberculosis. The value of this community-based approach was that the vulnerability of all age groups could be clearly identified. I sought to re-associate affected bones from individuals, based on their size, bone density and other physical traits. Eight children had characteristic bone modifications when they died. One was aged 3–5 years, two were 6–16 years, and five were 17–25 years when they died. There were minimally 18 adults, whose bones were more difficult to distinguish one from another. The extensive bone changes in some very young children showed that the pathogen was endemic within this community. It was puzzling that at least 6% of the skeletons in the ossuary showed bone modifications. Based on modern data, that suggested that everyone had lived with TB. That surely could not be.

Based on that study, I was approached by Mark Braun, a researcher at University of Indiana School of Medicine, who believed it might be possible to find DNA of the pathogen in affected bone. Mycobacteria like those causing tuberculosis are very robust, capable of surviving in the soil. Perhaps their DNA could be extracted. At his request, I sent him a small sample of bone from an affected vertebra. After literally years of work, applying the newly developed method of polymerase chain reaction (PCR), Braun managed to identify 123 base pairs of DNA that were diagnostic for TB (Braun, Collins Cook, & Pfeiffer, 1998). Such a tiny fragment of genetic material was big enough to confirm that it came from the *Mycobacterium tuberculosis* family, but nothing more. This and similar studies from other labs marked the beginning of molecular explorations of archeologically-derived bones from all over the world (Roberts & Buikstra, 2003).

There have been fits and starts, but gradually a coherent picture has emerged, one that not only highlights the lives of past peoples, but also describes the natural history of one of humanity's most feared killers. There are now automated, high-throughput systems for detecting pathogens like TB in archeological bone (Hubler et al., 2019). Studies of modern global diversity, combined with archeological studies, have shown that there are many strains of *Mycobacterium tuberculosis* affecting people and other animals. An early idea that a single human form came from the bovid form in Egypt a few thousand years ago has been rejected. That theory cannot explain the diversity of strains that have affected humans. The version of TB that was present in the Americas long before the era of European discovery was different from the European form. It may have come with the first migrants, crossing from Asia (Galagan, 2014), or it may be an independent mutation event in which the disease jumped from sea mammals to humans (Bos et al., 2014). The complete genome of the earliest American version of the disease hasn't yet been established, although this will likely come soon.

The contributions from bioarchaeological studies of ossuaries represent building blocks in a global enterprise to understand this family of pathogens. The fact that tuberculosis in humans can come in many different versions is relevant to the story of ancestral Huron-Wendat people. The disease that caused those puzzling bone modifications may have had unique characteristics. It may not have been as virulent, nor as debilitating, as the European version. This may help to explain why later exposure to European tuberculosis hit Indigenous communities so hard. We may not have all the answers, but this story illustrates that we cannot use analogy from European history to extrapolate about ancestral Huron-Wendat health and quality of life.

Ossuary communities: Collective osteobiographies

What can be learned about past peoples, when those few, precious instances of preserved ancestral remains are comingled? The more traditional approach of osteobiography, in which the individual is the centerpiece, is not feasible. In an ossuary, individual stories are co-mingled, but stories about the community may be discerned. The large numbers of people interred by each community make ossuaries very suitable for more population-oriented, lab-based approaches like stable isotopes and ancient DNA. Each of these story lines has been described in more detail elsewhere. My intent here is to illustrate the way different approaches can be woven together to create a collective osteobiography.

In discussions leading to a Memorandum of Understanding between the Huron-Wendat Nation and the University of Toronto (2011) (see Chapter 6), descendants agreed to embark on an ambitious program of research to learn new things about their ancestors. They authorized the curation of one tooth per person, plus small samples of disease-altered bone. The teeth and the bone tissue would be curated by the University of Toronto, on behalf of the descendants. This has led to a program of research that has posed and answered several questions about the ancestors' lives, from the 13th through 17th centuries. It has corrected past inaccuracies, corroborated evidence coming from artifacts, and provided an enhanced, more individualized picture of the past, broadly providing evidence of resilience.

A tooth can provide three kinds of tissue, each of which has unique qualities for chemical analysis (Fig. 5.3, Box 5.1). Enamel, which is almost completely inorganic, does not remodel—so it gives a snapshot of foods eaten while the crown of the tooth was forming. Because strontium sometimes stands in for calcium, we can learn about the bedrock geology in the region from which the food came. Dentine and bone have some similarities, although dentine is denser and remodels much less than bone. Both dentine and bone contain collagen, which provides information about the protein sources in the diet. Dentine will tell us about protein when the tooth was formed, bone will tell us about protein in the months just prior to death.

Within and among the three tissues are opportunities to explore several stable isotope systems. We studied three: carbon, nitrogen, and strontium. Much

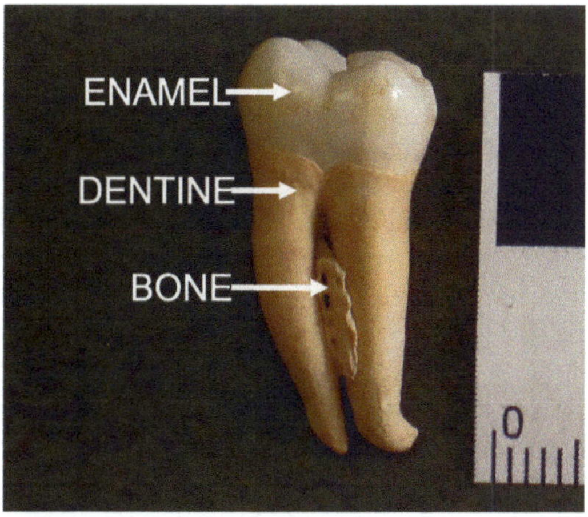

Fig. 5.3 The types of tissue that can be derived from a human molar; bone adhering to the roots is not always present. (Credit: S. Pfeiffer.)

Box 5.1 Isotope analyses from dental tissues

This table illustrates the qualities of three tissues—enamel, dentine and bone—with which archeological scientists can work. Each has unique traits. Enamel, which is almost completely inorganic, does not remodel—so it gives a snapshot of foods eaten while the crown of the tooth was forming.

Stable carbon isotope ratios reflect differences in plant metabolism, as determined within the earth's carbon cycle. Nitrogen isotope ratios reflect foods' positions in the food chain, also called the trophic level. These values are determined by a person's diet.

Because strontium sometimes stands in for calcium in the enamel of the forming tooth crown, we can use it to learn about the bedrock that underpins the soil, plants and water in the region. Generally, younger rocks (like chalk, for example) will have lower values than older rocks, like granite. The strontium value as measured in the enamel is set by the environment in which a child spends their early years. If that value differs from that of the environment in which a person was buried, it is evidence of mobility during the person's life.

Enamel and dentine capture information from the years during which the tooth was forming, while bone isotope values reflect the person's diet in the months prior to death. While many teeth may release from the jaw cleanly, with no adhering bone, the bifurcated molar roots often retain some bone trapped between them (see Fig. 5.3).

Continued

BOX 5.1 Isotope analyses from dental tissues—cont'd

Dentine and bone tissue have many similarities, although dentine is denser and remodels much less than bone. Both dentine and bone contain collagen, which provides information about the protein sources in the diet. Values from dentine tell us about types of protein in the food ingested while the tooth was forming. Values from bone tell us about types of protein in the months just prior to death. Note that none of the stable isotope dietary indicators will tell us whether there was enough food. They tell us about food types, not food sufficiency.

	Inorganic?	Carbon	Nitrogen	Strontium	Remodel?
Enamel	96%	✓		✓	NO
Dentine	75%	✓	✓		Slight
Bone	40–45%	✓	✓		Ongoing

dietary reconstruction has been done studying the carbon isotope ratios from bone collagen. The collagen in bone reflects mainly the protein in peoples' diets. Because maize is mainly a dietary source of carbohydrates, the proportion of maize in the diet needs to get quite high before it modifies the carbon ratio in collagen. Prior studies had shown substantial differences among ancestral Huron-Wendat communities, making it look like reliance on maize bounced around through time rather than trending up or down. We chose to measure the carbon values in the enamel, which does not favor protein during its formation.

The first step, after permission had been obtained and teeth had been set aside, was to confirm that a single tooth provides more than enough tissue to complete several analyses. Choosing ten teeth representing seven sites, we separated the root from the crown. Each root was explored for mtDNA, while the crowns became part of a larger study ($N=53$) of dietary isotopes. For the dietary study, we chose teeth where bits of bone adhered to the roots, since we needed to see how carbon isotope values from enamel and dentine lined up with those from bone collagen. All ten tooth roots yielded basic information about their mtDNA haplogroup (Pfeiffer et al., 2014). The dietary part of the project demonstrated the tighter group ranges when enamel is used instead of collagen for carbon isotope ratios (tracking maize consumption), and it demonstrated dramatic variability in nitrogen isotope values (tracking sources of protein) (Pfeiffer et al., 2014). The upshot of this rather modest paleogenomic study was what Northern Iroquoians fit well in the mitochondrial network of North American Indigenous peoples.

The upshot of the dietary work was that we needed to do a lot more work to understand the range of proteinaceous foods that Northern Iroquoians were relying on.

Dietary isotopes

The second, larger exploration of dietary isotopes expanded to 167 teeth from sixteen Northern Iroquoian sites. We also built a framework of food values, including 55 deer and 21 fish bones, all from archeological contexts. Our results show, even more clearly than the first study, a gradually increasing reliance on maize for food energy, with a significant up-tick in the 17th century (Pfeiffer, Sealy, Williamson, Needs-Howarth, & Lesage, 2016).

The expansion of the food web, providing data demonstrating dramatic variability in nitrogen isotope ratios among types of fish, helped to explain why the earlier studies of bone collagen had generated such a confusing picture. Fish are amazingly diverse and complex. For example, the bones of eels from different stages in their life cycle generated notably different nitrogen isotope ratios. A community relying on seasonally available catches translates into a broad range of nitrogen isotope ratios.

We then zeroed in on nitrogen isotope ratios measured from the ancestors' dentine. Because of the way that a tooth develops during infancy and early childhood, tiny slices of dentine can provide information about breastfeeding and weaning. We can see chemical signs of this because the baby is feeding off the mother's body, so is at a higher rung of the food chain. The enrichment is a direct reflection of the amount of the infant's food coming from breast milk. The first deciduous (baby) molar starts forming before birth and ends at about 2.5 years of age. A first permanent molar will have just begun laying down dentine at that point, ending years later (Fig. 5.4). Taken together, the first molars of the two dentitions form a continuous record of diet during about the first eight years of life. Summarizing our study of both deciduous and permanent teeth, the process of weaning began sometime between 8 and 18 months, and in all cases was complete by 3.5 years. Just as we saw with the carbon isotope results, the 17th century shows a different pattern. Children from these later sites show an earlier shift toward a diet without breast milk (Pfeiffer, Sealy, Williamson, Forrest, & Lesage, 2017). These two indicators of food insecurity—the increased proportion of carbohydrate in the diet and the earlier weaning of children—reflect the disruptions of the seventeenth century. Those difficult times affected the Wendat in ways that are similar to what is seen among people in regions of disruption and tumult today. Research on contemporary

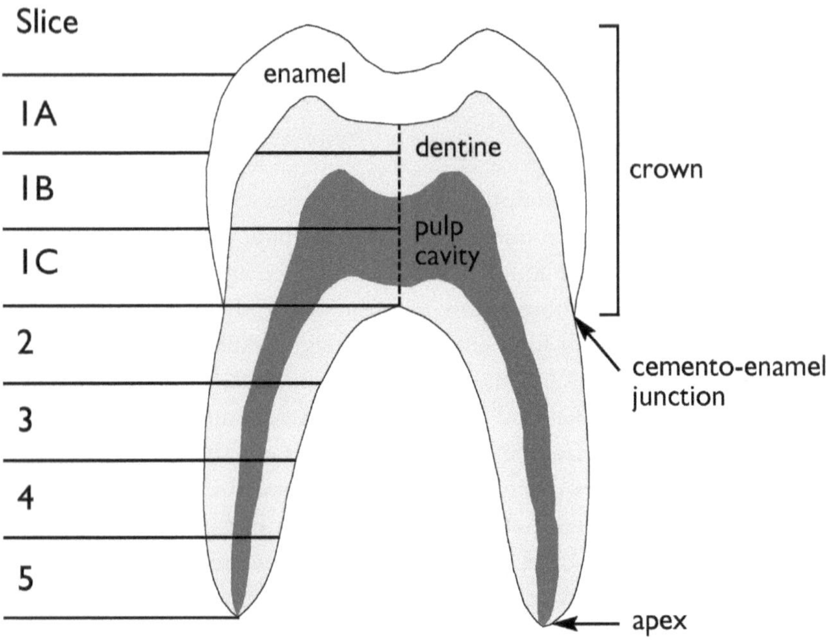

Fig. 5.4 Diagram illustrating one way dentine can be sampled, from first formed (1) to last (5), to study infant and childhood diet. Prepared by Neil Rusch. (Pfeiffer et al., 2017, fig. 2).

global communities has linked food insecurity to caregiver distress, childhood under-nutrition, and challenges to early childhood development (de Oliveira et al., 2020). Our studies confirm that these were tough times for the Huron-Wendat.

The co-mingled and broken-up nature of the ossuaries means that we mostly cannot know the sex of each skeleton, but a few adult lower jaws were intact enough to estimate biological sex from chin shape. When we look at the nitrogen isotope ratios of tooth roots (post weaning) from those jaws, we see a significant difference in dietary protein sources (Fig. 5.5). Boys had access to a higher proportion of low trophic level protein (presumably deer), and girls had access to a higher proportion of higher trophic level protein (presumably diverse types of fish). A possible explanation may be that at early ages, boys and girls were respectively introduced to their future gender-related roles in the community. Boys may have accompanied men outside the vicinity of the village to hunt deer. Many men, probably with boys' assistance, were needed to find wood for construction and maintenance of palisade-type fences and other structures, perhaps while clearing

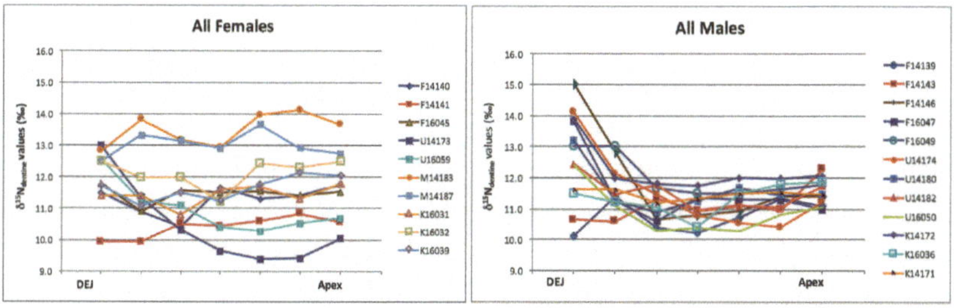

Fig. 5.5 Nitrogen isotope values measured in slices of permanent tooth dentine from mid-tooth to root tip; the sex differences suggest that boys and girls typically had different diets (Pfeiffer et al., 2017, fig. 5).

land to meet the ever-increasing extent of agricultural fields. Girls may have been introduced to the manufacture of pottery, collection of naturally occurring plants, and the planting, tending and harvesting horticultural crops. These tasks occurred closer to the villages. Villages' proximity to streams would have made some types of fish readily available (Pfeiffer et al., 2017).

Environmental isotopes

Moving to the strontium isotopes, as measured in the enamel, these required considerable background work because the environmental values for the region had not been determined. With the help of many archeologists and museum curators, we obtained enamel from deer and small mammals from secure archeological contexts. We measured their strontium values.

We found that the small mammal enamel values differ, depending on the major geological formation beneath the local soils. Human enamel is formed in infancy and early childhood. The human strontium value will match that of the small mammals from the place the person was buried, if the person spent her/his life in that region. If the person's value differs from the local signal, they moved some time after childhood.

Mostly, people's strontium isotopes indicate a lifetime within one geological zone. However, there were some deviations from that pattern. We could use maps of Ontario geology to surmise movement patterns. Most people buried in the Uxbridge ossuary spent their childhoods in the region to the south. One person buried in an ossuary in the more recent Georgian Bay cluster of sites shows an anomalous strontium value as well as the carbon isotope profile of someone whose childhood diet included little or no maize. That person appears to have been a hunter-gatherer from elsewhere, adopted into the group, buried within the ossuary (Pfeiffer et al., 2020).

Concluding remarks

The involvement of Huron-Wendat descendants and our discussion of results with that community has in every case enriched our interpretations of research results. I have encapsulated the projects here to illustrate the potential of collaborative, science-based research into the bones of ancestors, even in situations where they were excavated decades ago, even when it might be argued that they have already been studied in some fashion. Science does not remain static. The questions being asked can and do change. The methods for answering the questions improve. By setting aside tissue samples, different in type but no larger than would be sampled at an autopsy, substantial progress can be made toward giving the past a more human face.

References

Anderson, J. E. (1963). The people of Fairty: An osteological analysis of an Iroquois ossuary. In *Contributions to Anthropology, 1961-1962, Part 1* (pp. 28–129). Ottawa, Canada: Department of the Secretary of State. National Museum of Canada Bulletin 193.

Birch, J., & Williamson, R. F. (2013). *The mantle site: An archaeological history of an ancestral Wendat community*. Lanham, Maryland: AltaMira Press.

Bocquet-Appel, J., & Masset, C. (1982). Farewell to paleodemograqhy. *J. Hum. Evo., 11*, 321–333.

Bos, K. I., Harkins, K. M., Herbig, A., Coscolla, M., Weber, N., Comas, I., ... Krause, J. (2014). Pre-Columbian mycobacterial genomes reveal seals as a source of New World human tuberculosis. *Nature, 514*(7523), 494. https://doi.org/10.1038/nature13591.

Bourgeois-Pichat, J. (1951). Le mesure de la mortalité infantile. I. Principes et méthodes. *Population, 6*, 233–248.

Braun, M., Collins Cook, D., & Pfeiffer, S. (1998). DNA from *Mycobacterium tuberculosis* complex identified in North American, pre-Columbian human skeletal remains. *Journal of Archaeological Science, 25*, 271–277.

Buikstra, J. E. (Ed.). (1981). *Prehistoric tuberculosis in the Americas. Northwestern University Archaeological Program Scientific Paper No. 5.*. Evanston: Northwestern University Archaeological Program.

de Oliveira, K. H. D., de Almeida, G. M., Gubert, M. B., Moura, A. S., Spaniol, A. M., Hernandez, D. C., ... Buccini, G. (2020). Household food insecurity and early childhood development: Systematic review and meta-analysis. *Maternal and Child Nutrition, 16*(3). https://doi.org/10.1111/mcn.12967.

Galagan, J. E. (2014). Genomic insights into tuberculosis. *Nature Reviews Genetics, 15*(5), 307–320. https://doi.org/10.1038/nrg3664.

Hubler, R., Key, F. M., Warinner, C., Bos, K. I., Krause, J., & Herbig, A. (2019). HOPS: Automated detection and authentication of pathogen DNA in archaeological remains. *Genome Biology, 20*(1). https://doi.org/10.1186/s13059-019-1903-0.

Jackes, M. (1986). The mortality of Ontario archaeological populations. *Canadian Journal of Anthropology, 5*(2), 33–48.

Knowles, F. H. S. (1937). Physical anthropology of the Roebuck Iroquois with comparative data from other Indian tribes. *National Museum of Canada Bulletin, 87*, 1–76.

Lewis, M. E. (2007). *The bioarchaeology of children: Perspectives from biological and forensic anthropology*. Cambridge, UK: Cambridge University Press.

Pfeiffer, S. (1984). Paleopathology in an Iroquoian Ossuary, with special reference to tuberculosis. *American Journal of Physical Anthropology*, *65*(2), 181–189.

Pfeiffer, S., & Fairgrieve, S. (1994). Evidence from ossuaries: The effect of contact on the health of Iroquoians. In C. S. Larsen, & G. Milner (Eds.), *In the wake of contact: Biological responses to conquest* (pp. 47–61). New York: A.R. Liss.

Pfeiffer, S., Sealy, J., Williamson, R. F., Forrest, C., & Lesage, L. (2017). Patterns of weaning among ancestral Huron-Wendat communities, determined from nitrogen isotopes. *American Antiquity*, *82*(2), 244–261. https://doi.org/10.1017/aaq.2016.36.

Pfeiffer, S., Sealy, J. C., Williamson, R. F., Needs-Howarth, S., & Lesage, L. (2016). Maize, fish and deer: Investigating dietary staples among ancestral Huron-Wendat villages, as documented from tooth samples. *American Antiquity*, *81*(3), 515–532.

Pfeiffer, S., Willamson, R. F., Sealy, J. C., Smith, D. G., & Snow, M. H. (2014). Stable dietary isotopes and mtDNA from Woodland period southern Ontario people: Results from a tooth sampling protocol. *Journal of Archaeological Science*, *42*, 334–345. https://doi.org/10.1016/j.jas.2013.11.008.

Pfeiffer, S., Williamson, R. F., Newton, J., le Roux, P. J., Forrest, C., & Lesage, L. (2020). Population movements of the Huron-Wendat viewed through strontium isotope analysis. *Journal of Archaeological Science-Reports*, *33*(102466). https://doi.org/10.1016/j.jasrep.2020.102466.

Roberts, C. A., & Buikstra, J. E. (2003). *The bioarchaeology of tuberculosis: A global view on a reemerging disease*. Gainesville, Florida: University Press of Florida.

Seeman, E. R. (2011). *The Huron-Wendat feast of the dead: Indian-European encounters in early North America*. Baltimore Maryland: Johns Hopkins University Press.

Shook, B. A. S., & Smith, D. G. (2008). Using ancient mtDNA to reconstruct the population history of Northeastern North America. *American Journal of Physical Anthropology*, *137*(1), 14–29.

Thwaites, R. G. (1896–1901). *The Jesuit relations and allied documents*. Cleveland: Burrows Brothers Company.

Warrick, G. A. (2008). *A population history of the Huron-Petun, A.D. 500–1650*. Cambridge, U.K.: Cambridge University Press.

Williamson, R. F. (2014). The Archaeological history of the Wendat to A.D. 1651: An overview. *Ontario Archaeology*, *94*, 3–64.

Williamson, R. F., & Steiss, D. A. (2003). A history of Iroquoian multiple burial practice. In R. F. Williamson, & S. Pfeiffer (Eds.), *Bones of the ancestors: The archaeology and osteobiography of the Moatfield ossuary* (pp. 89–132). Ottawa: Canadian Museum of Civilization.

Williamson, R. F., Thomas, S. C., & MacDonald, R. I. (2003). The archaeology of the Moatfield village site. In R. F. Williamson, & S. Pfeiffer (Eds.), *Bones of the Ancestors: The Archaeology and osteobiography of the Moatfield Site* (pp. 19–88). Ottawa: Archaeological Survey of Canada.

CHAPTER 6
Collaboration and repatriation

Prologue

It was a brief encounter at one of those big institutional events. Someone important was about to speak, and I happened to be sitting near a member of my department, a social anthropologist. In my attempt at small talk, I told her that repatriation discussions with First Nations representatives were going well, including the idea that tissue samples might be retained for ongoing scientific research. At this, her face reddened. It seemed to me that her response was delivered through clenched teeth, though my memory may be playing tricks on me. She faced me down and said that the imposition of my scientific mindset onto First Nations persons was an affront to Indigenous knowledge and Indigenous ways. I replied that some Indigenous folks were genuinely interested in science as a way of knowing. Her retort, as she rose to change seats, was that I and my kind had imposed our Eurocentric views, forcing Indigenous people to fit in, and that science would never be part of their story. I have thought a lot about that assertion since that encounter. I think it is just as wrong to reify intellectual approaches as it is to consider dress styles, food ways and political parties as fixed in time. Even very traditionally oriented Indigenous people may choose to integrate scientific knowledge about their ancestors into the stories of their past and their hopes for their future.

Construction of a collaborative framework

The previous chapter reviewed how the idea of osteobiography can be applied using contemporary research methods, generating clearer and expanded information about ancestral lives. This chapter takes a different perspective, that of the people who hold the administrative responsibility for ancestral remains. Quite a bit has been written about the framework mandated by the federal law in the United States, NAGPRA (see Chapter 1). Most countries do not have federal laws mandating repatriation.

Osteobiographies
https://doi.org/10.1016/B978-0-12-823880-6.00007-0

Most managers of academic collections, and many museum curators, are more or less on their own. This chapter outlines the steps that were taken to accomplish the repatriation that I know best, the 2013 repatriation of Huron-Wendat ancestors. This action was broadly defined and mandated in a 2011 Memorandum of Understanding between the University of Toronto and the Huron-Wendat Nation. There followed soon thereafter a second Memorandum of Understanding between the Huron-Wendat Nation and the Ontario Heritage Trust (OHT) relating to the reburial on OHT land. These MOUs defined the repatriation plan and they established long-term, ongoing relationships among the three entities. I hope that our experience will be helpful to those who find themselves responsible for boxes of human bones, and those who suspect that their ancestors are, indeed, in someone's cupboard.

Preliminary steps

Canada has no federal legislation analogous to NAGPRA (first enacted in 1990). This may be, at least in part, because of effective outreach dating back to the 1970s in which archeologists and bioarchaeologists sought collaboration with descendant communities (Buikstra, 2006). The Canadian Association for Physical Anthropology (CAPA) endorsed a "Statement on the Excavation, Treatment, Analysis and Disposition of Human Skeletal Remains from Archaeological Sites in Canada" (Cybulski, Ossenberg, & Wade, 1979) that urged individual researchers as well as agencies to consult and communicate with local communities, including native band councils (the Canadian government refers to recognized Indigenous groups as bands). This is not to say that everything went smoothly henceforth, but at least the idea was out there, being discussed.

The focus of the CAPA statement and subsequent discussions was on discovery and excavation, not curation. A strong catalyst to think about the curated collections came as the Royal Ontario Museum (ROM) transferred the remains from the Ossossané ossuary (see Chapter 3) to Huron-Wendat descendants. The Ontario Heritage Act specifies that the licensed archeologist acts as steward of archeologically derived materials on behalf of the people of Ontario, so the ROM had the authority to take this action. The ROM owned the land from which the bones had been removed, so the bones could be placed there again. Ownership was transferred to the Huron-Wendat Nation of Wendake, Quebec, and the bones were reburied at the original location in 1999 (Kapches, 2010). While the repatriation was to be celebrated, it was distressing to scholars that this collection had been curated for half a century, yet very little research and resulting new knowledge had come from it, and now there could be none. What had been the point?

Meanwhile, ideas about how to manage the collections at the University of Toronto Department of Anthropology remained rather unfocused. They were housed on three campuses, and some collections had been moved many times during their many decades of curation. A framework for action was needed. Two Anthropology faculty members pressed ahead with the development of a departmental policy that would create a protocol for repatriation. The departmental executive committee approved the Anthropology Department Repatriation Policy in spring, 1999. This was about the same time that a request for the repatriation of one of the curated ossuaries was received. The request had come from an Indigenous person taking action as an individual. Hand wringing (figuratively speaking) ensued. In the summer of that year I moved from my prior institution, became a faculty member in the department and was made chair of the repatriation committee. This responsibility made sense, since my research had included study of some of the department's collections during the prior three decades.

The departmental policy, somewhat revised in 2020, is posted on the department's web site. In addition to addressing repatriation, it also firmly states that First Peoples skeletal material will not be used for teaching. This initiative predated federal ethics guidelines that were developed subsequently (most recent version: Government of Canada, 2018). The core function of the policy was, and continues to be, to facilitate the return of human remains and other culturally sensitive materials in Department of Anthropology collections to appropriate persons or groups, in an appropriate manner. An aspect of the procedure outlined in the policy is the recognized need for inventory, documentation and scientific inquiry (by which some destruction of tissue is implied), subject to approval "from the relevant First Nations groups." This latter part is newly established since the policy's creation and complies with the university's research ethics board protocols.

Getting back to the olden days of 1999, the Faculty of Arts & Science provided modest seed money for expert consultation. In 2002, the department requested and received a report prepared by curators J. Cybulski of the Canadian Museum of Civilization (now the Canadian Museum of History) and D. Ubelaker of the Smithsonian National Museum of Natural History. As we had requested, their guidance focused on how to document our collections, which originated with excavations undertaken in the 1940s through 1970s, and which at that point were not well documented or systematically housed. With this guidance in hand, we applied for funds to the Social Sciences and Humanities Research Council of Canada (SSHRC), and we were rejected. Documentation was not research, in their eyes. Later, in 2004, we made another application to federal research agencies, this time collaboratively with First Nations partners, the ROM, and a heritage consultancy, in the context of a special aboriginal initiatives competition. Once again, we were unsuccessful. Throughout the repatriation process, the financial burden for preparations fell to the institution.

Meanwhile, other developments

I think that big, old institutions are fascinating. Founded in 1827, now boasting over 2500 faculty and 60,000 students, the University of Toronto falls into that category. The quality and reputation of such an institution is built on the actions and contributions of many people, over a long time. This is mostly good, but inevitably there are some awkward moments. In 2003, 280 boxes of artifacts from 23 archeological sites, some of them ancestral Huron-Wendat, were removed from a campus tunnel (where they violated fire code regulations) and trucked to a landfill in Michigan (Globe & Mail, Caroline Alphonso, June 6). It was an unintentional destruction, but, understandably, there were repercussions. In response to the discussions that ensued, the President of the University of Toronto wrote letters to a number of Indigenous descendant groups, including the Huron-Wendat, pledging to do better at acknowledging and honoring the aboriginals who came before.

Much of southern Ontario was booming with construction, and consulting archeologists were very busy. Each time Indigenous human remains were encountered, an appropriate government-recognized band located in Ontario was consulted. This meant that the Six Nations of the Grand River (a Haudenosaunee community) were very busy, evaluating sites that had been created and used by ancestral Wendat, some dating from the period of intense conflict between the two Nations. After years of this, the Huron-Wendat decided to make their presence known, and what could be more straightforward than requesting that their ancestral remains be repatriated? Their legal representative asserted that U of T was legally vulnerable on the grounds of having disrespected the dead, and for having made empty promises. Whether these charges were valid or not, this approach was an attention-getter. Discussions between lawyers ensued.

By 2006, in response to many issues arising through land development activities, a consortium of three First Nations was established, called the Founding First Nations Circle (FFNC): Six Nations of the Grand River, Kawartha Lakes Anishnaabe, and Huron-Wendat of Wendake, Quebec. The concept was that the three bands would act as a secretariat when heritage issues arose. From the perspective of repatriation, this would have created an excellent approach. Each heritage collection might not have to be categorized as ancestral to a specific contemporary community. The FFNC designated an independent archeologist, Dr. Ron Williamson of Archeological Services Inc., to represent their specialist archeological interests.

The Anthropology department Repatriation Committee determined that the FFNC represented the descendants of ancestors represented by human remains from a substantial number of archeological sites. A smudging ceremony was performed (see Chapter 3). We set about documenting a broader set of collections and negotiating a Memorandum of Understanding with the FFNC. It was during this period that our proposal to retain tissue samples for long-term research purposes was tabled and discussed. The representatives took the idea back to their communities for consideration. There were precedents for the retention of one tooth per person. This approach was followed in 1997 when the Moatfield ossuary was removed, studied and reburied (see Chapter 4). Based on those results as well as others, we could describe the sorts of questions that could be addressed through scientific study of bone and tooth chemistry (see Chapter 5). Why had we focused on a tooth? It includes several types of biological tissue, its pulp cavity has been shown to protect endogenous DNA, and teeth that are shed during life are typically not imbued with spiritual meaning (tooth-fairies excepted). Since the removal of a tooth by a dentist is a familiar fact of life, the removal of an ancestral tooth seemed plausible. The discussion of this proposal uniquely illustrates the necessity for solid engagement across the table. We were asking our First Nations colleagues to think about approaches to learning that were unfamiliar to them. They, in turn, expected us to be transparent, reliable partners, and to construct interesting, researchable questions about real people who lived long ago, using methods that were minimally invasive.

Regrettably, and for reasons having nothing to do with U of T, the FFNC collaborators abandoned their consortium approach in 2008. In 2010, the Huron-Wendat Nation contacted the University, suggesting that we proceed with a Memorandum of Understanding between our two institutions. This was affirmed by the departmental Repatriation Committee and University leadership. Building on the principles within the draft plan made

Fig. 6.1 Grand Chief Konrad Sioui signing of the Memorandum of Understanding between University of Toronto and the Huron-Wendat Nation. *(Credit: S. Pfeiffer.)*

with the FFNC, the MOU was completed in 2011, signed in Wendake, Québec (Fig. 6.1) (Pfeiffer & Lesage, 2014).

Preparing for repatriation

Identify the locus of leadership

As the process unfolded, it became apparent that organizational structure was our friend. At both the institutional and departmental levels, we relied on a relatively narrow responsibility tree. The University, which is itself a government institution, made it clear that it would make agreements with other government-recognized institutions. This was important as other individuals and groups who self-identify as First Nations pressed their cases. It was up to our partner, the Huron-Wendat Nation, to interact with them. So too, in the Department of Anthropology, a team made up of me, a staff person and a postdoc were the figurative "bag holders." The post-doc and I are both trained in bioarchaeology. To our good fortune, the Huron-Wendat Nation asked Ron Williamson to continue his role. He and his office team took charge of documenting the associated objects that had been excavated with the human remains and were to be returned and reburied. Many other people played important roles. For example, administrators supported our request for dedicated space in a discrete location and provided funding. Nevertheless, the preparatory work required long term commitment to a relatively narrow plan and streamlined decision making, executed with a fair amount of discretion.

Document pragmatically

While discussions were underway, there was much to be done if repatriation was to be executed smoothly and appropriately. The existence of the Anthropology Department's Policy on Repatriation was of foundational importance. As months went by, it became apparent that the consultants' report was less useful. Our colleagues had provided guidance in principle, absent specific details of sample sizes and preservation quality. They recommended a full expert assessment of each bone and tooth, following the *Standards for Data Collection from Human Remains* (Buikstra & Ubelaker, 1994). That is a document that focuses on individual skeletons that are not comingled and broken into small pieces. Although we embarked on that approach briefly, it soon became clear that it wasn't feasible. We were preparing over 1700 individuals; many bones were in very small pieces, many teeth were "loose"—out of their sockets. For some collections, the only maximum length measurements would be of a few hand and foot bones, from adults of undetermined sex. For other collections, some metric and nonmetric data were available in previously written theses, publications, and unpublished reports. Our efforts were better spent hunting for those reports, verifying and compiling details regarding the sites, excavations and subsequent studies.

Pragmatism was needed, too, when we addressed the assignment of archeological collections to contemporary descendant communities. The Anthropology Department houses many research collections. Which of the skeletal collections should be seen as ancestral to the Huron-Wendat Nation? Although humans create their identities through both biological and cultural lenses, this was primarily a question of cultural affiliation, not biology. All archeological sites dating to before European involvement were created by people genetically defined as Indigenous Americans. With some effort, and some destruction of tissue, one could genetically confirm this for each skeleton. There are two reasons why this was neither plausible nor helpful for us. For one thing, because a study of ancient DNA would be classified as "human subjects research," the test would have to be okayed by either the subject or a relative speaking for the subject. There is no straightforward work-around for a test designed to see who the ancient person was related to. But in this case, ancient DNA was not likely to provide a clear answer to the question because a community is a culturally defined, ethnic group. As noted by the Jesuit relations, a village could and did adopt people from other groups. There had not been rigid social barriers to mate choice, nor any physical barriers like oceans or mountain ranges. The tools to define identity would need to be archeological. The antiquity and cultural affiliations of archeological collections from this time and place are most accurately determined from a combination of the artifacts reported, the pottery styles, site features (like the presence of a relatively deep ossuary, or long house dimensions), and European trade goods in the later decades. The archeologist representing the Huron-Wendat, Ron Williamson, played a central role in helping us make these judgments. It becomes progressively more difficult to associate sites from earlier times with modern cultural groups because of the dynamic, shifting nature of human culture.

Call in the loans

Bringing all the bones into one place was a big task. Portions of collections had been loaned to other institutions, with varying levels of documentation. The memories of long-time staffers were crucial. If, for example, a doctoral student had done her thesis research on an ossuary collection, she might have taken some portion of the collection across the country to her subsequent faculty position, in order to continue her research. Ah, but that junior scholar now had retired, her department had been re-organized and moved to a new building, and no one had any records that went back that far. You get the idea. For one site that was said to have a minimum of 132 persons included, we were able to locate about half. We discussed frustrations like these with our First Nations colleagues, and sincerely appreciated their forbearance. Locating the objects that had been found with the bones was particularly difficult, especially if they had been loaned. The current practice of storing burial gifts with the deceased person is a relatively new approach to curation. The various utilitarian objects and adornments reported in site reports were often missing. Since we often had no firm sense of their dimensions, nor how well preserved they were, we could not say if they had just disintegrated (eroded iron objects, for example), broken to bits (small glass beads) or been moved to storage with inadequate labeling. In every instance of calling in the loans, we had to rely on the willingness of the possessor to comply with the repatriation request. Some were sympathetic to the initiative, others were not. Our institution sometimes needed to make transport arrangements and pay shipping costs. In due course, all the bones and associated objects that we could locate were returned to the University of Toronto.

Some collections included extensive burial gifts. At the behest of the Huron-Wendat, Williamson and a committed team at Archeological Services Inc. spent literally months working to document the artifacts that had never been properly examined. They brought in blacksmiths with expertise in 17th century techniques to assess metal objects, as well as archaeozoologists and experts in glass beads . Their analyses and the related analytical data were enhanced by professional quality photographs of each object.

Build a vocabulary

Even the focus of our activities—repatriation—needed some thought in the context of our discussions and documents. Is "repatriation" a gendered word? After all, words like patrimony and partriarchy are clearly masculine. Should we be talking about repatriating and rematriating, referring to returns to fatherland and/or motherland? Some etymological digging suggests that this is not necessary. The root of the word, patria, is Greek for "one's own country." In Greek, this noun is not gendered. We decided that the word repatriation is gender neutral.

The vocabulary I had developed as a bioarchaeologist was one that emphasized the distance between the researcher and the subject. This was meant to convey objectivity. It was important that discussions with descendant groups made progress toward mutual understanding and respect. Their non-verbal cues provided quick feedback that my language was not taking the right tone. So, we talked about it. I was encouraged to engage with the human remains as those of deceased human beings, to be considered multidimensionally. Sometimes the adjustments to language addressed the recognition of personhood of the deceased. In other times, they clarified the nature of proposed activities, like the distinction between documentation and research (Table 6.1). As is often the case with language, once my references to "individuals" shifted to talk of men, women, infants, and so on, my mental template for the work also underwent a subtle shift. This is a work in progress. For example, descendants would prefer that the bits of tissue retained for subsequent research not be called "samples," but it is tough to think of a better term.

Show good faith

On a sunny day in May 2010, the newly elected Grand Chief of the Huron-Wendat Nation was coming to visit the Department of Anthropology, with his advisors and lawyer. I was nervous. I admit to standing at the glass façade of the Anthropology building, watching for their cars. After parking, the

Table 6.1 Terminology and definitions developed during repatriation discussions.

Term	Definition/comments
Repatriation	Transfer of responsibility for human remains and associated artifacts
Descendants	A. The university works with First Nations governments recognized by Canadian provincial and national governments. B. Assessment is based on geographic, scientific, archeological, material culture and contextual evidence
Documentation	The observations and measurements needed to characterize human remains; includes various non-destructive approaches
Research	Scientific approaches that modify or destroy human tissue, including dating, histology, chemistry, and genetic analyses
Collection, individual, male, female, subadult	Terms to be avoided, as they would not be applied to family members
Person, man, woman, infant, juvenile, child	Preferred terms

Modified from Pfeiffer, S., & Lesage, L. (2014). The repatriation of Wendat ancestors/Le rapatriement des ancêtres Wendat, 2013. Canadian Journal of Archaeology, 38, 5–26.

group assembled, then clustered at the street corner for a photo. Why? Because the Anthropology building is at the intersection of Russell Street (re-named Ursula Franklin Street in 2020) and HURON Street. Indeed, the Hurons had come to Huron Street, to visit their ancestors.

The meeting was constructive, as various ideas for repatriation were debated. Much remained unclear. At this point, the question of where the ancestors might be reburied remained unanswered. There would be more meetings of smaller groups, but this large gathering had stimulated one particularly productive idea. The Anthropology building has a large first floor lobby, which in 2010 was open space with benches and blank walls. The ROM had given us four surplus display cases, sitting empty in the center. Looking out onto Huron Street, why not use the space to inform people about all the communities who had called the locale home, long before U of T? A collaboration of the Department with the ROM (which provided curatorial staff), Archeological Services Inc., the Huron-Wendat Museum at Wendake, Quebec, and the Woodland Cultural Center at Brantford, Ontario, went smoothly. The grand opening of the permanent display,

"Uncovering our Early Past: First Nations in Toronto" took place on May 16, 2012 (University of Toronto, 2012). The display cabinets hold artifacts and other materials from Departmental archeological collections, the walls provide timelines and basic background information. It was one of the first initiatives at U of T done to acknowledge Indigenous stewardship of the lands on which the institution is built.

Prepare the ancestors for their next journey

Once the ancestral Huron-Wendat collections were identified, we needed to prepare them for their next journey. All extraneous material was removed from the stored human remains. All paper, plastic, foil, string and other material was removed. Many of the bones had labels written on their surfaces with India ink. These labels could not be removed without damaging the tissue beneath, and so they stayed. All the human remains and burial gifts bound for reburial were placed into new cardboard "bankers boxes" that were constructed so that no labeling was visible on the outside. We weighed each one, taking care to keep each filled box light enough for community participation at the reburial. The boxes would be flattened and reburied as well, so that even the bone dust would be returned to the earth.

We set aside left mandibular molars as we proceeded. This could be either a deciduous or a permanent tooth, depending on the age of the deceased person. First molars were preferred. With a well-trained eye, the deciduous and permanent first molars could be distinguished from among the loose teeth. From among mandibles and mandibular fragments, many of the adults had lost their first molars to decay prior to death, so sometimes a second molar was set aside instead. If a tooth came from a child's mandible, measurements of the jaw and the developmental status of the other teeth were noted, providing a more precise age-at-death to associate with that tooth. If a tooth was taken from an adult mandible, measurements and information about bone shape were recorded, since men and women tend to have different jaw lines. This gave us some information about the probable sex of the person. Each tooth was placed in its own small bag, with the name of the collection and any other details noted on the outside. The same process was followed with the bits of disease-altered bone.

A venue for the reburial had been secured. The land from which one of the ancestral ossuaries had come was owned by the provincial government. The OHT had agreed to request re-zoning for that parcel so that it would be a legally recognized cemetery in perpetuity. Plans were made for creation of a pit that was large enough for all the skeletal remains, and safe enough for community members to participate in the return of their ancestors to the earth.

Repatriation and after

September 2013: Specially chartered busses traveled the 800 km, bringing dozens of Huron-Wendat participants from Wendake, Quebec, to Toronto, Ontario for a three-day celebration. On the first day, the rooms holding the prepared boxes were cleansed with a smudging of sweet grass. After securing one building entrance to assure no onlookers, the boxes were loaded onto trucks and conveyed to the final resting place. Huron-Wendat volunteers would stay with the boxes until the final burial. Meanwhile, a small conference was held in which interested researchers, descendants and their representatives discussed the potential for future discoveries from scientific research on the retained tissue samples. This gathering focused on describing possibilities and building linkages.

On the second day, a large university hall was filled to capacity for an event celebrating the Huron-Wendat presence in Ontario (Fig. 6.2). Arrangements had been made for simultaneous translation, since most of our Huron-Wendat guests were unilingual French speakers, and most of our presenters were unilingual English speakers. Presentations described diverse aspects of archaeology, history and bioarchaeology, in an atmosphere that was both scholarly and celebratory. In keeping with traditions of generosity, the lunch was superb.

On the third day, the chartered busses delivered the Huron-Wendat to the reburial locale, about 45 km northwest of Toronto. All the ancestors would be re-interred at what had been the Kleinburg ossuary, now re-named in honor of a past Huron-Wendat leader, Thonnakona. A deep trench with reinforced side walls and an entry with a shallow pitch had been prepared. The boxes were removed from the trucks and the reburial proceeded. This event was attended solely by descendants and their guests, who represented other First Nations groups. No visual or audio documentation, and no unauthorized persons were allowed. As they explained it, the prior day was akin to a public funeral, whereas the committal was just for the family. It was interesting and a little disheartening that these decisions, plus the absence of visuals, acted as significant disincentives to media representatives who had initially been quite engaged. There was very little news coverage as over 1700 skeletons were reburied. When the busses returned

Fig. 6.2 The Great Hall of Hart House, University of Toronto, on the occasion of A Celebration of the Huron-Wendat Nation in Ontario, September 2013. *(Credit: Archeological Services Inc.)*

to downtown Toronto that evening, we all joined together for a big party. Once again, the food was superb.

Feeling quite good about it all, I quickly realized that there was still a lot of work to do. All the retained teeth and small samples of disease-altered bone needed to be archived, using a coding system that would allow us to keep track of the retained materials. When the Huron-Wendat authorized research, there would need to be documentation that tracked both the loan and the return of the remaining tissue. Each fragment of disease-altered bone was photographically documented so that interested researchers could view images, available on request. A coding system was designed that gives each sample (tooth or bone) a unique identifier, labeled by tissue type and collection. Additional notes are linked to those identifiers.

A suitably sized, lockable cabinet was purchased for the samples. It was placed in a secure room with limited access. Loan/return forms were developed that would document access to the tissue samples, subsequent to the approval of proposals by descendants. Once again, I was learning about something new to me: the centrality of a material transfer agreement (MTA). This is a "contract that governs the transfer of tangible research materials between two organizations, when the recipient intends to use it for her or his own research purposes. The MTA defines the rights of the provider and the rights and obligations of the recipient with respect to the materials and any progeny, derivatives, or modifications" (Wikipedia, June 30, 2020). All this infrastructure was designed to support the University's intent to act as a repository on behalf of the Huron-Wendat Nation, well into the future.

Things to not do

There were a few things that we did during the decades preceding repatriation that were well meant, but not helpful. Human remains in curation are frequently stored in cardboard boxes. If the boxes are free of vermin, and if the bones have been in those boxes for a long time, it is of little value to re-box them, and even less use to replace packing material (often newspaper) with acid-free paper. These actions may be of use in long-term curatorial contexts, but the preferences of the receiving group should be the first priority. Until the receiving group indicates how they wish to receive and transport their ancestors, curators should resist the temptation to cosmetically alter the collection.

Another well-meant initiative is the recruitment of undergraduate volunteers to work on documentation. In the early 1990s, a sessional lecturer assigned the bones of one ossuary to undergraduates enrolled in a course. Each student had a category of data to collect, as part of their coursework. Their submitted papers were meant to expedite the documentation of the collection. The problems with such a plan include the variable quality of the information in those reports and the absence of written permission from each student to use the data for subsequent purposes. All those hours of data collection were of no subsequent use. So, what about recruiting undergraduate volunteers to help research staff? This was seen by our First Nations colleagues as a continuation of a longstanding practice of non-Indigenous learners benefiting from access to the remains of their ancestors. They argued that privileged students had had "fun with the bones" long enough. They were comfortable with the involvement of well trained, paid professional staff, including graduate students. I once listened to a discussion in which a professor asked an Indigenous person, "How will we train the next generation of students if they cannot learn from working with these collections?" The reply was, "That is not my problem." Indeed, it isn't. How would you feel, if your ancestors were handled in this way, without your knowledge or consent?

References

Buikstra, J. E. (2006). Repatriation and bioarchaeology: Challenges and opportunities. In J. E. Buikstra, & L. A. Beck (Eds.), *Bioarchaeology: The contextual analysis of human remains* (pp. 389–415). Amsterdam: Academic Press.

Buikstra, J. E., & Ubelaker, D. H. (1994). *Standards for data collection from human skeletal remains* (1st ed.). Fayetteville: Arkansas Archeological Survey Research Series #44.

Government of Canada. (2018). *Tri-council policy statement: Ethical conduct for research involving humans*. Ottawa: Secretariat on Responsible Conduct of Research.

Cybulski, J., Ossenberg, N. S., & Wade, W. D. (1979). Statement on the excavation, treatment, analysis and disposition of human skeletal remains from archaeological sites in Canada. *Canadian Review of Physical Anthropology*, *1*(1), 32–36.

Kapches, M. (2010). Ossossané ossuary: The circle closes. *Archaeology of Eastern North America*, *38*, 1–15.

Pfeiffer, S., & Lesage, L. (2014). The repatriation of Wendat ancestors/Le rapatriement des ancêtres Wendat, 2013. *Canadian Journal of Archaeology*, *38*, 5–26.

University of Toronto (2012). Uncovering our Past: First Nations in Toronto [Retrieved from https://www.youtube.com/watch?v=iUagC9yAAXo] (May 16, 2012).

CHAPTER 7

Framework for osteobiographies: Eastern and Southern Africa

Prologue

I first visited South Africa in 1992. I was going to give a brief talk at a conference in Cape Town. I was using the trip as an opportunity to learn more about the region's archeological community and its curated collections of excavated human remains. As a mid-career academic, I'd felt badly stalled. The drive behind my Ph.D. thesis—an interest in the lives of past hunter-gatherers—had been sidelined for many other projects. I was captivated by those whose lives and livelihoods were ultimately constrained by the natural world. I wanted to get back to that earlier focus, but where in the world might I find the factors needed for impactful scholarship? A research program—a set of related research questions that generates significant new knowledge—takes some careful design. Given my training and interests, the ideal context would have evidence of communities successfully exploiting the landscape for a long, long time, without recourse to domesticated herds or crops. The archeological records of that exploitation should be methodologically sound. It would be wonderful if the soils were conducive to excellent bone preservation. I wanted to ask questions about how people managed under such circumstances, which, after all, represent most of our species' time on earth. As the subsequent chapters will show, I found all that and more in Africa.

I had tried to prepare myself for the trip, reading about the geography and recent history of the country. Once the flights were behind me, I was pleased to see that the cities had many signs in English (one of South Africa's ten official languages). The people I met were wonderfully gracious. That was good, because the left-side driving made me dependent on others for transport. As I tried to absorb everything passing by my car window, I noted a pattern. Goods were touted as being "for Africa." There were Shoes for Africa, Toys for Africa, Food for Africa, Trucks for Africa, and so forth. What was that about? A host explained to me that the term "for Africa" represents something that is endless, vast, abundant. Okay, it means a vendor has lots of shoes. I got it. Even within Africa, there was a sense that this

Osteobiographies
https://doi.org/10.1016/B978-0-12-823880-6.00008-2

continent could not be compartmentalized. Even Africans think of Africa as almost immeasurable in its scope and diversity.

The continent of Africa is home to 54 countries, according to the United Nations (Fig. 7.1). The number of languages spoken is somewhere between 1250 and over 3000, according to Wikipedia. I have visited only nine of the 54 countries and I speak no local language. I am no expert on Africa. Nevertheless, I have spent a lot of time there, and a lot of time thinking about the past lives and resulting genetic diversity of the billion-plus people who call some part of Africa home.

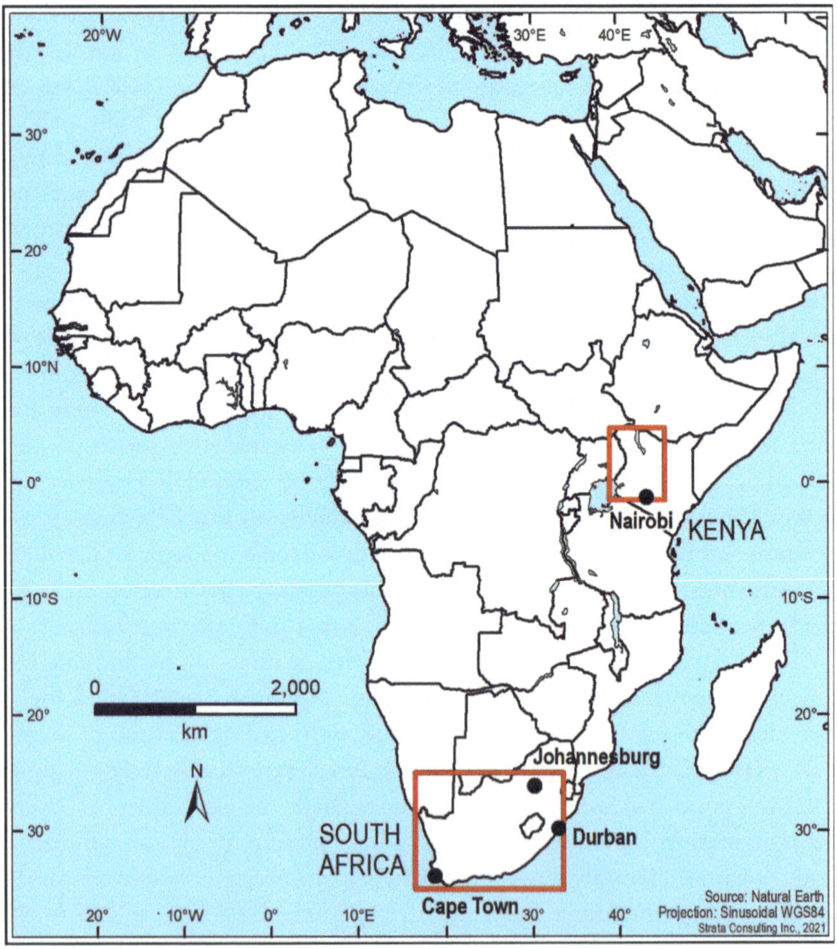

Fig. 7.1 A map of Africa; rectangles highlight the regions of particular interest in subsequent chapters.

The lure of Africa

Why do people from other continents travel to Africa? Top tourism picks are game parks and sites of fossil discoveries. People aspire to visit eastern Africa to stand on the edge of Olduvai Gorge where Louis and Mary Leakey made their impressive discoveries from the 1930s onward, and to marvel at their determination. They want to see the rugged landscapes of the African Rift, where fossils like the famous partial skeleton called "Lucy" (found in 1974) wait to be discovered. In southern Africa, they may want to explore the karstic caves from which the bones of Australopithecines and their relatives have been extracted for almost 100 years. This region now claims the designation "Cradle of Humankind." Fortunately for the tourists, there is good game viewing in those regions, too. Based on the skeletal record, it is certain that Africa played a central role in the evolution of our lineages prior to two million years ago. I have thrilled to visit the fossil sites and the museums dedicated to their interpretation. I delight in revisiting the images. Fossil evidence indicates that ancestral forms spread from Africa to the rest of the world several times, the most recent being an ancestor that looked a lot like us: anatomically modern *Homo sapiens* (Wood, 2019). Surely this hotbed of human evolution has not been static since various ancestral forms departed Africa, heading for Eurasia and beyond. Africa is where many mysteries remain. As David Reich has noted, "Focusing on Africa as the place where our species originated, while it might seem to highlight the importance of Africa, paradoxically does Africa a disservice by drawing attention away from the question of how populations that remained in Africa got to be the way they are today" (2018: 208).

My research interests align with a focus on Africa as the home to our own lineage, and on the people of Africa who carry humanity's most ancient genome. This is a statement that I could have only vaguely framed back in 1992. It is built on discoveries that have been made since that date, arising from collaborations of palaeoanthropology and genomics. Anthropologists knew that African populations were diverse not just in their languages and cultures, but also in their anatomy. This diversity was mostly ignored by European and American scholars of the 19th and early 20th centuries. Although those scholars had a lot of respect for the ancient Egyptians and a few other north African groups, they simply placed the peoples of sub-Saharan Africa at the base of their racial hierarchies and quibbled about which rungs on the ladder of evolution belonged to the "Caucasoids" and the "Mongoloids" (Gould, 1981). As biological sciences progressed, and goals became more geographically ambitious, more quantitative attempts were made to expand the human "organizational chart"—often characterized as a tree with many branches, but with some branches more important than others. By the mid-20th century, bones, teeth and other aspects of bodily form, like hair and noses, were assessed as proxies for genes in a more global context (an example of textbooks of the time is Kelso, 1970).

A historically significant example of the racial typologies of that era is *The Origin of Races* (Coon, 1962). The author identified five races of "man" (the term meant to subsume all genders): Australoid, Mongoloid, Caucasoid, Capoid and Congoid (Fig. 7.2). The last two in the list represent sub-Saharan Africans. The Congoid group included all darkly pigmented sub-Saharan Africans, including pygmies. The term Capoid was the group now identified as San or Khoe-San. Despite many conceptual flaws in the basic premise of distinct races, the separation of the two African categories was apt. The Khoe-San genome is now recognized as being the most distinctive of all living peoples (Nielsen et al., 2017; Schlebusch et al., 2017). The label "Khoe-San" (spelled variously) is an academic construct that includes many self-determined modern communities who do not use that identifier for themselves, but who have a deep, shared ancestry (see Box 9.1). The Khoe-San are also known as the Bushmen. Their success as foragers, exploiting available resources from the land, is legendary. Today, communities live interspersed throughout Southern Africa. They speak, or recent generations spoke, distinctive Khoisan (yes, a variant spelling) languages, which rely heavily on "click" sounds. They are also known for their distinctive and successful approaches to hunting and gathering, as well as their petite body frames.

The scenarios and interpretations linked to all the old categories, including Capoid, were embedded in biased, hierarchical thinking. The Capoids, that is, the Khoe-San or Bushmen, were thought to have originated in North Africa with considerable Caucasoid input, hence their lighter skin tones. Their diminutive body size was discussed in the context of archeological evidence: "…there are full-sized skulls of un-reduced ancestral Bushmen. Similar populations persisted in the northern Transvaal until after A.D.1000, although without doubt the mysterious shrinking, fetalizing syndrome that produced the modern Bushmen had by that time begun elsewhere." (Coon, 1965:92). As subsequent chapters will show, the myth of the recent diminution of the Bushmen was still in play decades later.

The racial characterization illustrated by works like Coon's was justly criticized on both ideological and biological grounds. In a landmark paper

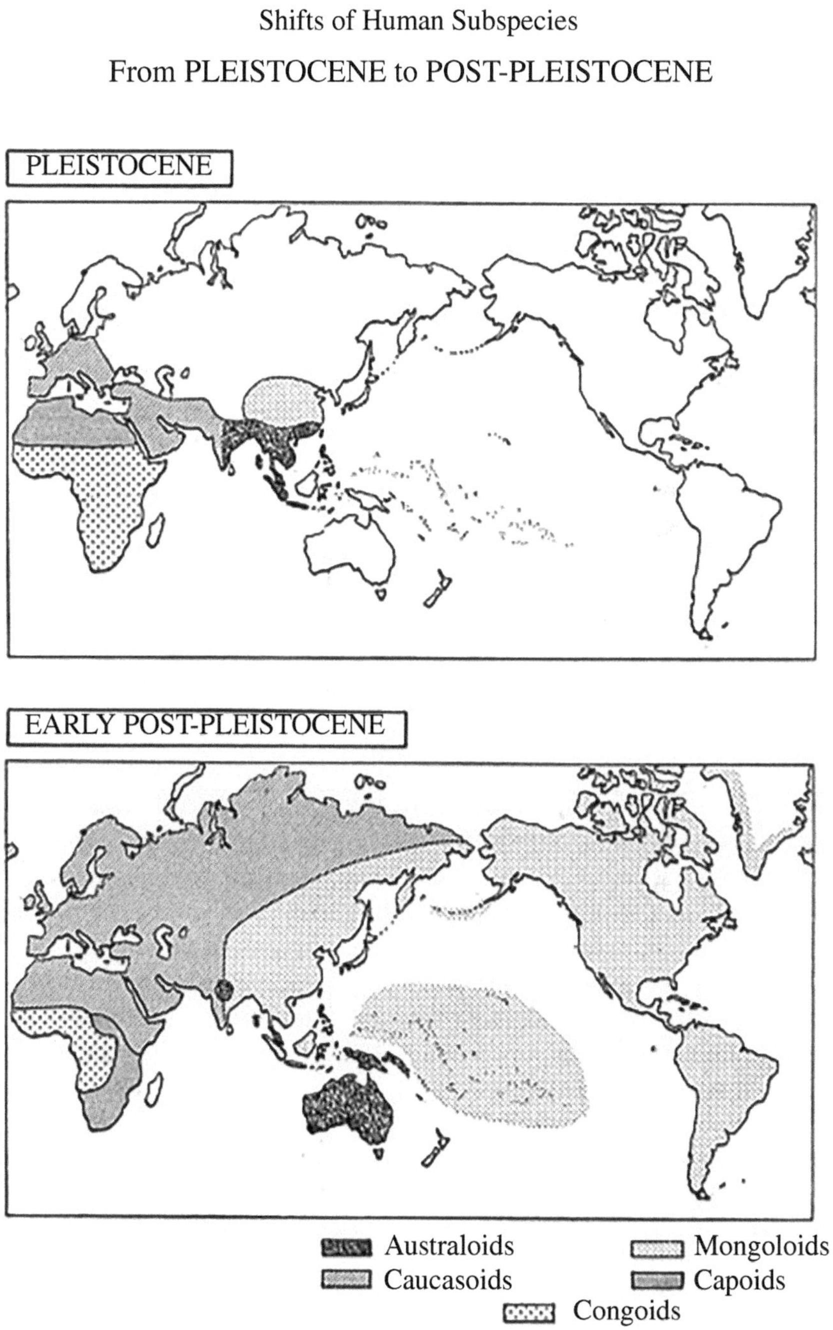

Fig. 7.2 A map from Coon's book, *The Origin of Races*. Map 13 (1962:659). Wikipedia.

that appeared soon after Coon's book, Richard Lewontin used the available evidence of variation in protein types in blood to demonstrate that most genetic variation (80–85%) is seen within local geographic groups, leaving only the remaining 15–20% for traditional "race" groups (Lewontin, 1972). The socioeconomic damage caused by racism was based on deeply flawed biological assumptions. The academic interest in characterizing diversity through assessment of how people looked diminished dramatically in subsequent decades. This history helps to explain why biological anthropologists closely scrutinize and often critique the interpretations made when new genomic evidence for human diversity is published (Saini, 2019). Once burnt, twice shy.

Modern humans: What role for Africa?

Since researchers set out to characterize the entire genome of living humans at the turn of the 21st century, the pace of discovery has been remarkable. Characterization of the patterning in the DNA of living humans provides the framework for interpretation of the DNA extracted from their ancient forebears. The first steps in the study of ancient DNA focused on mitochondrial DNA (mtDNA). There is a lot of it. It is present in every nucleated cell and its structure is very simple, compared to that of nuclear DNA. The study of mtDNA led to the theory that our ancestors left Africa rather recently, and that all non-Africans descended from relatively few breeders. We were said to have descended from a mitochondrial Eve. This idea was to change as new information was added from ancient nuclear DNA. The time frame was going to shift, and the story was to become more complex. The characterization of the Neanderthal genome, followed by the characterization of the genome of the barely-there Denisovans (hardly any fossils, but well preserved genes within them), led to the observation that living non-Africans carry small bits of Neanderthal and Denisovan DNA. Africans and people of African descent do not carry small bits of those particular extinct distant relatives. It is now quite common to see articles in the popular press about "our" Neanderthal component, as though a bit of Neanderthal and/or Denisovan is the shared heritage of all living populations. It is not. One aspect of the mitochondrial Eve story that has been verified is that modern sub-Saharan African human DNA—both mitochondrial and nuclear—is more variable than that of other humans. While current evidence is not entirely clear about the role Africa played in the story of modern human ancestry (Bergström, Stringer, Hajdinjak, Scerri, & Skoglund, 2021), fossils and genomic evidence from the continent will continue to be very important.

Archaeology and palaeogenomics are forging ahead to characterize "Who We Are and How We Got Here" (as per Reich, 2018), giving many European and Asian ethnic groups clear, concise origin stories built on syntheses of heritage sites and the genes of ancient skeletons. Discoveries have not kept the same pace with respect to African ethnicities. By the end of 2018, only 85 ancient human genomes (or parts thereof) had been published from the continent of Africa, relative to 3500 from Eurasia (Brunson & Reich, 2019). Challenges to the expansion of the African story include poor tissue preservation (diagenesis, see Box 7.1), cultural habits of burial that make it difficult to locate skeletons, and, of course, research resources. Neither excavation nor palaeogenomic research comes cheap, and African settings often present substantial challenges.

BOX 7.1 The link between palaeogenomics and osteobiography

From the earliest observations of human genetic differences—even before the discovery of the structure of deoxyribonucleic acid (DNA) in 1953—researchers attempted to link living people to skeletons that might be their ancestors. Most commonly, the skull shapes of the living were used as templates to categorize the skulls of the past. The tool kit has changed, but the questions about population structure are still of interest (Table B7.1).

The first published segment of human aDNA, achieved using PCR, was published in 1985. Reports of fragments of mtDNA and Y chromosomes drew attention, but one prominent researcher has noted that "genetic evidence before

Table B7.1 Terms, abbreviations and definitions.

Terms	Abbreviations	Definitions
Deoxyribonucleic acid	DNA	Hereditary material made up of four chemical bases, organized in pairs; contains biological instructions, also called genetic code
Mitochondrial DNA	mtDNA	Short DNA chains, maternally transmitted, found in all nucleated cells
Ancient DNA	aDNA	DNA extracted and sequenced from an organism long after death, usually "ancient" meaning more than 100 years old (Brunson & Reich, 2019)

Continued

BOX 7.1 The link between palaeogenomics and osteobiography—cont'd

Table B7.1 Terms, abbreviations and definitions—cont'd

Terms	Abbreviations	Definitions
Polymerase chain reaction	PCR	A technique that replicates specifically targeted fragments of aDNA
Next generation sequencing	NGS	Also called high-throughput sequencing (HTS); a number of technologies that detect and sequence millions of aDNA fragments in parallel, preparatory to bioinformatic analysis
Single-nucleotide polymorphism	SNP	A single base-pair difference in the DNA sequence of an individual; from 1 M to 20 M SNPs from each genome are compared
Bioinformatics		A field of computational science focusing on the analysis of sequences of biological molecules
Principal component analysis	PCA	A technique often used to visualize aDNA variation, based on relationships of clusters (Coutinho, Vincente, & Schlebusch, 2020)
Paleogenomics		The analysis of aDNA from a sufficient number of locations to obtain statistically meaningful information (Brunson & Reich, 2019)
Archaeogenomics		Palaeogenomic research on nonhuman animals and plants from archeological sites, analyzed in relation to that context (Brunson & Reich, 2019)
Diagenesis		The transformation of components within mineralized collagen (bone and dentine) following deposition/burial, compromising preservation (Kendall, Eriksen, Kontopoulos, Collins, & Turner-Walker, 2018). Hot climate, for example, accelerates chemical reactions that degrade DNA

BOX 7.1 The link between palaeogenomics and osteobiography—cont'd

around 2009 was mostly incidental to studies of the human past…, a poor handmaiden to the main business of archaeology" (Reich, 2018:22).

The combination of NGS technology and bioinformatics creates the basis of paleogenomics, which in turn can extend into investigations of the genomes of past animals and plants, termed archaeogenomics (Brunson & Reich, 2019; Coutinho et al., 2020). Fragments of aDNA from each person or organism are assayed for millions of SNPs, and patterns are then compared. Dramatic discoveries have come mainly from geographic regions with cool, dry climates. The tissues holding the aDNA are less diagenetically altered—better preserved—if their burial environment has protected them from moisture and warm temperatures, factors that encourage microorganisms and chemical reactions with the surrounding matrix. While any part of a well-preserved skeleton may produce a good yield of aDNA, some parts of the body preserve aDNA particularly well. Preferred sampling sites are teeth with intact pulp chambers and bone from a part of the skull called the petrous pyramid, a part of the temporal bone. It surrounds the cochlea—the inner ear. Its formation early in development, its density, and the near absence of subsequent remodeling contribute to a high level of DNA preservation (Hansen et al., 2017; Pinhasi et al., 2015). This sampling site opens up opportunities for aDNA extraction in situations where the dentine of tooth roots is absent, or has been diagenetically compromised. On the other hand, well preserved teeth are better sources for the aDNA of diseases, because more blood flows through them.

Genomic research relies on large sample sizes for its analytical power, and palaeogenomic research is built on the same model. This has led to a scramble for target tissues, gathered directly from excavations and from museum collections. There are concerns about the management of this destructive testing (Fox & Hawks, 2019; Palsdottir, Blauer, Rannamae, Boessenkool, & Hallsson, 2019), and initiatives are underway to better manage it (Prendergast & Sawchuk, 2018). Some noteworthy palaeogenomic discoveries have focused on single skeletons, like the resolution of the conflict over the linkage with Indigenous descendants of the American skeleton known as Kennewick Man (Rasmussen et al., 2015). Rarely is the connection of between an ancient genome and the sampled skeleton's osteobiography pursued, but it can be enlightening. For example, the partial skeleton of a little boy who lived 2000 years ago along the southeast African coast yielded well-preserved aDNA (Schlebusch et al., 2017). His bones show that he was anemic, which was very unusual for that community. We suggested that schistosomiasis, which is still endemic in the region today, contributed to his premature death (Pfeiffer, Harrington, & Lombard, 2019). Because his genomic material is so well preserved, a search of this aDNA sample for pathogens could uncover evidence of factors that challenged his health. Combining bioarchaeological and

Continued

BOX 7.1 The link between palaeogenomics and osteobiography—cont'd

palaeogenomic research can bring a sense of the lived experiences of persons behind the aDNA samples. As the field of palaeogenomics matures, researchers are working toward a more holistic approach to encounters with humans of the past.

Brunson, K., & Reich, D. (2019). The Promise of Paleogenomics Beyond Our Own Species. *Trends in Genetics, 35*(5), 319–329. https://doi.org/10.1016/j.tig.2019.02.006

Coutinho, A., Vincente, M., & Schlebusch, C. M. (2020). DNA is the key to unlocking our ancient African past. *The Biochemist, 42*(1), 12–17.

Fox, K., & Hawks, J. (2019). Use ancient remains more wisely. *Nature, 572,* 581–583.

Hansen, H. B., Damgaard, P. B., Margaryan, A., Stenderup, J., Lynnerup, N., Willerslev, E., & Allentoft, M. E. (2017). Comparing ancient DNA preservation in petrous bone and tooth cementum. *PLoS One 12*(e0170940).

Kendall, C., Eriksen, A. M. H., Kontopoulos, I., Collins, M. J., & Turner-Walker, G. (2018). Diagenesis of archeological bone and tooth. *Paleogeography Palaeoclimatology Palaeoecology, 491*, 21–37. https://doi.org/10.1016/j.palaeo.2017.11.041

Palsdottir, A. H., Blauer, A., Rannamae, E., Boessenkool, S., & Hallsson, J. H. (2019). Not a limitless resource: Ethics and guidelines for destructive sampling of archaeofaunal remains. *Royal Society Open Science, 6*(10). https://doi.org/10.1098/rsos.191059

Pfeiffer, S., Harrington, L., & Lombard, M. (2019). The people behind the samples: Biographical features of past hunter-gatherers from KwaZulu-Natal who yielded aDNA. *International Journal of Palaeopathology, 24*, 158–164. https://doi.org/10.1016/j.ijpp.2018.10.008

Pinhasi, R., Fernandes, D., Sirak, K., Novak, M., Connell, S., Alpaslan-Roodenberg, S., … Hofreiter, M. (2015). Optimal Ancient DNA Yields from the Inner Ear Part of the Human Petrous Bone. *PLoS ONE, 10*(6). https://doi.org/10.1371/journal.pone.0129102

Prendergast, M. E., & Sawchuk, E. (2018). Boots on the ground in Africa's ancient DNA "revolution": Archeological perspectives on ethics and best practices. *Antiquity, 92*(363), 803–815. https://doi.org/10.15184/aqy.2018.70

Rasmussen, M., Sikora, M., Albrechtsen, A., Korneliussen, T. S., Moreno-Mayar, J. V., Poznik, G. D., … Willerslev, E. (2015). The ancestry and affiliations of Kennewick Man. *Nature, 523*(7561), 455-U159. https://doi.org/10.1038/nature14625

Reich, D. (2018). *Who We Are and How We Got Here: Ancient DNA and the New Science of the Human Past*. New York: Pantheon Books.

Schlebusch, C. M., Malmström, H., Günther, T., Sjödin, P., Coutinho, A., Edlund, H., … Jakobsson, M. (2017). Southern African ancient genomes estimate modern human divergence to 350,000 to 260,000 years ago. *Science, 358*, 652–655.

"Population structure" is the term for a population's organization of genetic information. That pattern reflects the combined effects of mutation, recombination, natural selection and other evolutionary processes. The initial attempts at reconstructing prehistoric African population structure were based on relatively scant information compared to parallel efforts in Europe, Asia and the Americas (Skoglund et al., 2017). Nevertheless, some of the early ideas seem to be holding true. Studies consistently show that the genome of the Khoe-San people of southern Africa is the "deepest," the most complex, representing one branch of the earliest population divergence in *Homo sapiens*. The Khoe-San are on one branch and everyone else is on the other. Paleogenomic studies persistently detect low levels of Khoe-San-affiliated ancestry among ancient eastern Africans associated with various non-Khoe-San cultural contexts (Wang et al., 2020). The ancestors of the Khoe-San may have lived throughout much of sub-Saharan Africa at around the time that anatomically modern *Homo sapiens* was departing Africa, heading off to see the world. The earliest known African skeletal evidence of anatomically modern *Homo sapiens* comes from the northern coast of the continent, in what is now Morocco (Hublin et al., 2017). Those bones are dated to about 300,000 years ago. They show no particular anatomical similarities to Khoe-San skeletal features. The puzzle pieces are so few that one cannot feel confident about any imaginative reconstruction of human population structure from that era.

Osteobiographies of past Africans

The divergence date for Khoe-San people and everybody else is estimated at 200–300,000 years ago. It is based on the nuclear DNA mutation rate, which is also an estimate, so it is always expressed as a range. The continued existence of a lineage for such long a time may be attributed in part to chance, but it is unlikely that "chance" tells the whole tale. Elements of geographic isolation and human ingenuity may also contribute to the story. Surely the persistence of this lineage for so very long justifies the pursuit of osteobiographic methods to better understand their success. A fascination with the Khoe-San story has motivated much of my research. Like the Huron-Wendat in North America, there is history and ethnography against which one can gauge the archeological evidence. There are significant stories to be told.

When balanced against all of sub-Saharan African diversity, the world of Khoe-San foragers represents an important but small piece. There are untold stories of other foraging peoples, of pastoralists, crop producers and urbanites. Compared to many other parts of the world, these people and their stories remain mostly unexplored. In the coming chapters I will offer a few glimpses, illustrating their value, helping in a small way to construct a time when scholars know as much about the past peoples of Africa as they do about the past peoples of the Northern Hemisphere. The toolkits of archeological science, bioarchaeology and palaeogenomics have been used to discover who built Stonehenge and how they did it. I look forward to a day when we have more detailed information about the accomplishments of the people who built Great Zimbabwe, Jenne Jenno, Aksum, and the many other impressive, ancient cultural landmarks of sub-Saharan Africa.

References

Bergström, A., Stringer, C., Hajdinjak, M., Scerri, E. M. L., & Skoglund, P. (2021). Origins of modern human ancestry. *Nature (London)*, *590*(7845), 229–237. https://doi.org/10.1038/s41586-021-03244-5.

Brunson, K., & Reich, D. (2019). The promise of Paleogenomics beyond our own species. *Trends in Genetics*, *35*(5), 319–329. https://doi.org/10.1016/j.tig.2019.02.006.

Coon, C. S. (1962). *The origin of races*. New York: Alfred A. Knopf.

Coon, C. S. (1965). *The living races of man*. New York: Alfred A. Knopf.

Gould, S. J. (1981). *The mismeasure of man*. New York: W.W. Norton.

Hublin, J. J., Ben-Ncer, A., Bailey, S. E., Freidline, S. E., Neubauer, S., Skinner, M. M., … Gunz, P. (2017). New fossils from Jebel Irhoud, Morocco and the pan-African origin of *Homo sapiens*. *Nature*, *546*(7657), 289. https://doi.org/10.1038/nature22336.

Kelso, A. J. (1970). *Physical anthropology: An introduction*. Lippincott.

Lewontin, R. (1972). The apportionment of human diversity. *Evolutionary Biology*, *6*, 391–398. https://doi.org/10.1007/978-1-4684-9063-3_14.

Nielsen, R., Akey, J. M., Jakobsson, M., Pritchard, J. K., Tishkoff, S., & Willerslev, E. (2017). Tracing the peopling of the world through genomics. *Nature*, *541*(7637), 302–310. https://doi.org/10.1038/nature21347.

Reich, D. (2018). *Who we are and how we got here: Ancient DNA and the new science of the human past*. New York: Pantheon Books.

Saini, A. (2019). *Superior: The return of race science*. Boston: Beacon Press.

Schlebusch, C. M., Malmström, H., Günther, T., Sjödin, P., Coutinho, A., Edlund, H., … Jakobsson, M. (2017). Southern African ancient genomes estimate modern human divergence to 350,000 to 260,000 years ago. *Science*, *358*, 652–655.

Skoglund, P., Thompson, J. C., Prendergast, M. E., Mittnik, A., Sirak, K., Hajdinjak, M., … Reich, D. (2017). Reconstructing prehistoric African population structure. *Cell*, *171*(1), 59. https://doi.org/10.1016/j.cell.2017.08.049.

Wang, K., Goldstein, S., Bleasdale, M., Clist, B., Bostoen, K., Bakwa-Lufu, P., … Schiffels, S. (2020). Ancient genomes reveal complex patterns of population movement, interaction, and replacement in sub-Saharan Africa. *Science Advances*, *6*(24). https://doi.org/10.1126/sciadv.aaz0183. eaaz0183.

Wood, B. (2019). *Human evolution: A very short introduction* (2nd ed.). Oxford, UK: Oxford University Press.

CHAPTER 8

Lake Turkana: An intriguing osteobiography

Prologue

The market town of Lodwar, northwest Kenya, is the capital of Turkana County. It is about 500 km north of Nairobi, 600 km northeast of Kampala, Uganda, 1700 km straight west from Mogadishu, Somalia. The airstrip at Lodwar is memorable for the large jet engine abandoned in the weeds at one end of the runway. The air services are crucial to development agencies supporting the water and food relief programs often needed by the people of this arid region, and to archeologists and palaeoanthropologists headed for Lake Turkana. Kenya's rather new constitution (ratified in 2010) specifies that the country's heritage resources shall be held regionally. This reverses a long history of everything being shipped to the Kenya National Museum, a globally important curation repository in Nairobi. Thanks to that constitution and the determination of Richard Leakey, a growing catalogue of important fossils and artifacts can be found at the end of a long, very bumpy road leading from Lodwar to the Turkana Basin Institute (TBI).

The TBI is a marvel of plumbing, electricity and internet connectivity on the banks of the Turkwel River. It is part of a network of fixed assets in Nairobi, Ileret (on the other side of Lake Turkana) and Turkwel. It is authorized by the Government of Kenya to serve as a repository for the archeological and paleontological heritage of Lake Turkana region. Managed by Stony Brook University (Long Island, NY) in close collaboration with the Leakey family, it is a site of year-round research and field schools. The research is organized into broad chronological bins. It was my good fortune to be included in the research of the Later Prehistory of West Turkana unit (LPWT) for a number of years. Indeed, I guess I still am, since my LPWT colleagues and I continue to work on projects arising from our discoveries there. I just don't fly to Lodwar anymore.

Osteobiographies
https://doi.org/10.1016/B978-0-12-823880-6.00009-4

The Lake Turkana Basin

Lake Turkana, formerly Lake Rudolf (after an Austrian prince, until Kenyan independence) is known locally as Anam Ka'alakol, the sea of many fish (Fig. 8.1). It sits in a cleft known as the Turkana-Omo Basin in the Great, or Eastern, Rift. The famous north-south African Rift Zone is known to students of human evolution as the site of many important fossil discoveries. Fossils of Paranthropus—the hominin with the huge teeth—and early *Homo* are important to the story of our origins (Wood & Leakey, 2011). The

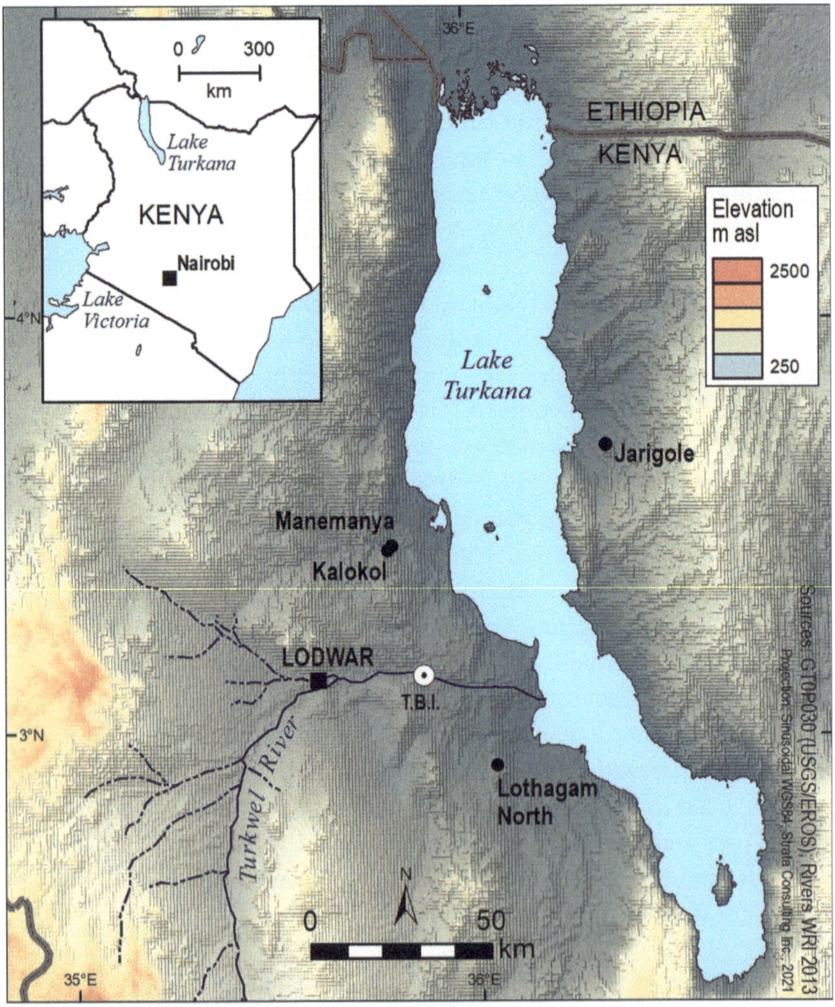

Fig. 8.1 Map of the region discussed in this chapter.

ancient sites, as old as 4 million years ago, can be dated by study of the volcanic layers that were exposed by the tectonic activity that formed the Rift. Early evidence of anatomically modern *Homo sapiens* remains ephemeral, so far. The lake's palaeoanthropological importance has led to detailed studies of its geology and ecology through time. The lake used to be a lot bigger than it is now. These days it continues to shrink, but it remains a major feature on the Eastern African landscape.

Today's sere landscape is one of dust devils waltzing among camels and goats as they browse the acacia trees. This is a relatively new phase, climatically speaking. The archeological documentation of modern humans exploiting the region dates back to the African Humid Period (AHP), which began around 15,000 BP and continued to about 5000 BP. In those days people were hunter-gatherer-fishers. Current aDNA studies (Chapter 7) suggest that these folks were related to the ancestors of today's Khoe-San genome. With lake levels high, there was connectivity across a broad swath of Africa, within which mobile groups of people would have exploited readily available resources. Lest we think that all was peaceful, 10,000 year old skeletons discovered at the Turkana basin site of Nataruk appear to have been left where they fell, with some showing signs of interpersonal violence (Lahr et al., 2016). What a jarring reminder of humans' propensity to compete aggressively.

By about 5000 years ago, as the climate commenced a prolonged drying period, archeological evidence suggests that people introduced a new subsistence component, supplementing their hunting, gathering and fishing with the herding of cattle, goats and sheep. It is not clear from archeological evidence whether these were newly arrived people from elsewhere, or locals trying something new. With time, the pastoralist lifeway became well established and is known archeologically as the Pastoral Neolithic (PN, 3000–1200BP). The transition preceding the PN is the context of the osteobiographic tale to come. Just as in the North American hunter-gatherer context described in Chapter 4, evidence of where people lived is very scarce. They appear to have been highly mobile, and there are questions about whether this was logistic mobility, residential mobility, or some combination of the two. Ethnographic information from modern pastoralists describe mixed strategies as people do their best to maintain their herds and exploit other resources. As in North America, observations of modern groups cannot be applied directly to the distant past.

The mobile hunter-gatherers described in Chapter 4 appear to have returned to special places, perhaps annually, to bury their dead. They created cemeteries that included some fully intact skeletons (primary burials) and some cremations (one form of secondary burial). About 5000 years ago, at locales around Lake Turkana, people created cemeteries of a different kind. While the structure of these lakeside cemeteries is unique, it is likely that they functioned like other cemeteries. They symbolize an intent to honor and stay connected to the people whose lives were important to the community.

The pillar sites

At a handful of locales around Lake Turkana, there appear large chunks of columnar basalt, far from where they naturally occur, positioned vertically through placement by human designers. They appear in groups. Sometimes they form circles, other times lines, and sometimes the design of their placement is obscure (Fig. 8.2). The people who now live in the region associate them with prior people of a long-ago time. Called ng'amoratunga, they are said to be dancing people turned to stone by a deity. At several of the sites, smooth, rounded basalt cobbles have been used to pave large circles. They, too, have been carried to the chosen site.

The first archeological explorations of the pillar sites were at the site of Jarigole, on the east side of Lake Turkana. Within the top meter of stone and soil, excavators found tiny fragments of human bones and teeth. They

Fig. 8.2 The pillar site at Kalokol, Turkana County, northwest Kenya, with an acacia thorn tree for scale. The site is visible from a main road. Passers-by visit and place small stones on the columnar basalt pillars. *(Photo: S. Pfeiffer.)*

suggested that the sites might have held fragments of ancestors, being ossuaries composed of secondary burials. While archeologists normally seek to explore a feature to its full depth, all manner of logistical problems intervened so that whatever lay more deeply remained unknown. The LPWT took on the question, focusing on pillar sites on the west side of the lake. Multiyear excavations (2009–2014) at Lothagam North determined that well beneath the first meter there were intact, primary burials. Based on the 36 skeletons found in a single unit that reached bedrock, we estimated that the complete feature might be the final resting place for several hundred people (Hildebrand et al., 2018). More limited excavations were done at other west-side pillar sites, and Jarigole was revisited. Sure enough, beneath that first meter the excavators encountered complete skeletons, buried soon after death.

The discovery that pillar sites are cemeteries was exciting. The excavated skeletons and their adornments contributed new information to the story of prior lakeside peoples' lives. It was accompanied by another discovery, that all the pillar sites explored so far—both sides of the lake—were constructed during a very narrow window in time. Radiocarbon dates from the sites are all from a period of about 5000 to 4300 years ago (Grillo & Hildebrand, 2013; E.A. Hildebrand et al., 2018; E.A. Sawchuk et al., 2019). This was a time and place about which very little had been known, except that the lake levels were dropping. Now there were many concrete discoveries that needed explanation.

Every generalization about the burials found at pillar sites seems to have an exception. The burials found so far are mostly adults, but there are some children and infants. Most burials are primary, but there are some extra elements that signal secondary burials. Most bodies were positioned with tightly flexed arms and legs, heads turned in diverse directions as though they had been tightly wrapped and enclosed in a shroud before placement. However, some bodies seem to have been interred together. The angles of the multiple bodies at Lothagam North suggest a placement of bodies one upon another in an open pit. Large stones were placed upon, and around, the bodies. Most items accompanying the bodies are clearly adornments: beads, bangles and rings, plackets, earrings and such. Most people seem to have been dressed in their finery. The surviving raw materials are mainly colorful types of stone and hippo ivory, although there are also novel materials like a head covering woven from at over 400 gerbil teeth and another item featuring over 40 hyrax teeth. The distinctively incised, unglazed pottery is present as fragments, not complete vessels. There is very little evidence of food waste, like the bones of fish or terrestrial animals, nor is there much evidence of post holes, ochre and hearths.

Artifacts link the pillar sites with habitation sites thought to represent pastoralists. Near the base of the Lothagam North main excavated unit was a burial with a stone bowl carved with a zoomorphic face resembling a cow. All evidence to date suggests that the pillar sites were created by pastoralists, although this mortuary tradition is not seen in the subsequent Pastoral Neolithic. It appears that toward the end of the African Humid Period, pastoralism was replacing hunting/fishing/gathering and people were organizing themselves into large groups, at least when the construction of a pillar site was called for. The planning, digging and hauling all suggest a coordinated collective effort. A large number of burials interred together, like at Lothagam North, suggests either an open pit for a long time or a very big population contributing their recently deceased family members. The Lake Turkana shoreline pillar sites could have served as fixed anchors for mobile communities who were creating new strategies to survive and thrive in a shifting climatic landscape. Evidence for their housing, their toolkit, their garbage, their geographic extent is all scant. What can we learn about them from the unique window provided by a few of their burials?

A significant challenge to osteobiography at the pillar sites is diagenesis, the transformation of mineralized tissues after deposition (see Box 7.1). All biological tissues undergo deterioration once they are buried. Soft tissues decay rapidly, but the speed and extent of changes to mineralized tissues like bone, dentine and enamel varies with the environment. Preservation of the bones and teeth within the pillar sites was very poor indeed. Being encased in a matrix of large pebbles and boulders for 5000 years had a number of effects. Their collective weight crushed the bones, of course. The placement of large stones (too heavy for one person) onto each body seemed to be part of the burial protocol. The accumulation of stones also hindered normal drainage so that moisture pooled around the organic pockets created by each burial. This generated a relatively active organic matrix in which the collagen in the bone and dentine deteriorated, thanks to microbial and chemical factors. The enamel, lacking an organic component, normally preserves well. However, once the dentine has deteriorated, the enamel crowns of the teeth can collapse, for want of internal support. As we excavated, the bones and teeth first appeared as badly broken and crushed, but otherwise normal in structure. Once the strong, hot sunlight dried their surface, they delaminated—that is, they fell apart. The loss of the organic (collagen) component left the mineral component of the bone unsupported. Very frustrating.

Much of the bone had to be stabilized with consolidant (Butvar-98, thinned with acetone). The teeth had their own protocol, since they provide crucial information about the person's age, and other traits through stable isotope explorations. Jaws were encased in foil and "excavated" in the lab.

Manemanya

The pillar sites represent an enigmatic archeological context. Everyone involved in their exploration seeks to contribute new knowledge, the better to understand this little piece of the human past. I am going to describe in considerable detail a single burial from one of the pillar sites, to illustrate how the methods of osteobiography can play a part.

On the modern landscape, the pillar site of Manemanya is much less imposing than Lothagam North. It was discovered by surveying specifically for chunks of columnar basalt found far from where they would be expected. A radiocarbon date from the initial exploration indicated a date within the range of the other pillar sites, toward the more recent end (Hildebrand & Grillo, 2012). Initial exploration had identified some human bone, so a team laid out one meter-square excavation unit in 2012. Five days had been budgeted, to determine the structure of the site by creating a cross-sectional profile. Working through the first meter, we found a few bits of bone and a few ostrich eggshell beads, but further down, we encountered more and more beads, then some skull bone. Once you encounter skull bone, normal excavation practice is to "pedestal it." That means that you figure out the skull's orientation from the suture lines and begin to remove the overburden in the direction you expect the skeleton to be positioned, assuming an intact, primary, extended or flexed burial. That approach can work well when the burial is in soil or sand, but it just wouldn't work at Manemanya. The skull had been crushed by a large rock, and indeed, there were large rocks covering and surrounding the body. We also knew, from our work at Lothagam North, that the bone would not withstand prolonged exposure to the intense Turkana sun. We proceeded to remove overburden within our square, and, to our good fortune, the extent of the square encompassed the position of the body (Figs. 8.3 and 8.4).

The bioarchaeology of the Manemanya burial

There are three levels of context that together tell us a story about this person: the biological features of the person, the interment environment and the items included, many as personal adornment.

Fig. 8.3 The author excavating and Elizabeth Sawchuk assisting, Manemanya, 2012. Note the dense matrix of rounded stones, which continued to a sandy soil base, seen beneath the burial's legs in Fig. 8.4. *(Photo: S. Pfeiffer.)*

Fig. 8.4 The partially exposed burial at Manemanya. Black arrow points to lower jaw. The person was lain on her/his back, with arms extended, and flexed legs drawn to the right. The torso is beneath a mass of beads. The head was sharply slumped onto the right shoulder. *(Photo: S. Pfeiffer.)*

The person: The complete permanent dentition and the completed growth of the vertebrae and long bones made it clear that the person died in adulthood. The roots of the third molars were completed as well, another indication of completed development. One fragment of a rib end looks like that of a young adult, and there is very little wear to the chewing surfaces of the teeth. We saw slight evidence of osteoarthritis on the articular ends of two ribs, but other joints, like the mandibular condyles and a few well-preserved hand bones, show no osteoarthritic lipping. All the indicators we could assess indicate a rather youthful adult, but again the features we would have liked to assess were not preserved. To our great frustration, the pelvic bones were quite damaged, but we noted shape characteristics on fragments that suggested a female hip shape. Also, the chin is not prominent, the brow region shows no prominence. These features suggest a female sex. All the bones that we could observe were quite delicately built, but the sex estimate remains tentative, especially since the skull shows some very unusual features. Very provisionally, we decided that the skeleton represents a young woman.

An estimated maximum femur length taken from the damaged upper leg bone while it was in the burial position can be used to generate a stature estimate of 170–180 cm (Lundy & Feldesman, 1989), which is relatively tall. Estimates of maximum length taken from fragmented bones are tricky. The estimated length of a humerus generates an estimated stature of only about 160 cm. A reasonable stance is that the person's stature was not unusually tall or short.

Moving on to health indicators, there is no clear evidence of cranial bone response to anemia and the teeth don't show developmental defects. The frontal bone retains a bit of the metopic suture: a developmental feature that normally fuses before the end of the first year of life but is sometimes retained without consequence. The facial bones are asymmetrical, with an unusual bony spur along the edge of the left cheek bone but none on the right side. The sides of the lower jaw are also imbalanced, with right and left sides shaped differently. The wear on the teeth, while not extensive, is sufficient to indicate more chewing and less calculus buildup on the right side than on the left. The skull vault is abnormally shaped, but it had spent 5000 years beneath a boulder, so it is hard to feel confident about the person's original traits in this regard. Moving toward the back of the skull, there are several ossicles (separate centers of ossification along the suture lines) (Fig. 8.5). Taken together, the facial asymmetries and the ossicles suggest an abnormality. Using palaeopathology's differential diagnosis approach, we have suggested that the person may have had an unresolved case of muscular torticollis (wry-neck) (Sawchuk et al., 2019). The clinical literature says that this condition is normally transient. That was not the case here. Perhaps the developmental asymmetries were linked to a neurological trauma, or some other soft tissue trauma or abnormality. Most cranial vault growth occurs before mid-childhood, so the presence of the sutural bones indicates that

Fig. 8.5 A view of the back of the warped, partially reconstructed Manemanya skull, showing the extra bones along the sagittal and lambdoid suture lines and a rather porous surface texture. *(Photo: S. Pfeiffer.)*

the condition was present from a very young age. The limb bones, hands and feet appear normal, to the extent that we could assess them in their deteriorated state.

The interment: It became apparent as we excavated that a thick layer of formerly strung ostrich eggshell beads covered the person's torso. Estimating the number from the weight from a counted sample of a few hundred, there were at least 10,000 beads that appear to have been thrown onto the body after it was laid to rest. Although the strands of beads extended to either side of the body, they did not extend beneath the body. In other words, they were not worn by the deceased; they were placed or tossed atop the body. The strings of beads covered the torso and pooled between the arms and the torso. Beneath the body, on the flat, rock-free soil surface, there was evidence of a small fire. We observed charcoal and ash, but the soil had not been modified by intense heat.

Garb and adornment: (Fig. 8.6) In the ear regions on both sides were small clusters of extremely small ostrich eggshell beads, 26 in total, that may have been strung as earrings. Around the neck was at least one close necklace of five calabash/gourd-shaped stone beads. There were also curved animal teeth and seed-type beads in that region, maybe from a second string or

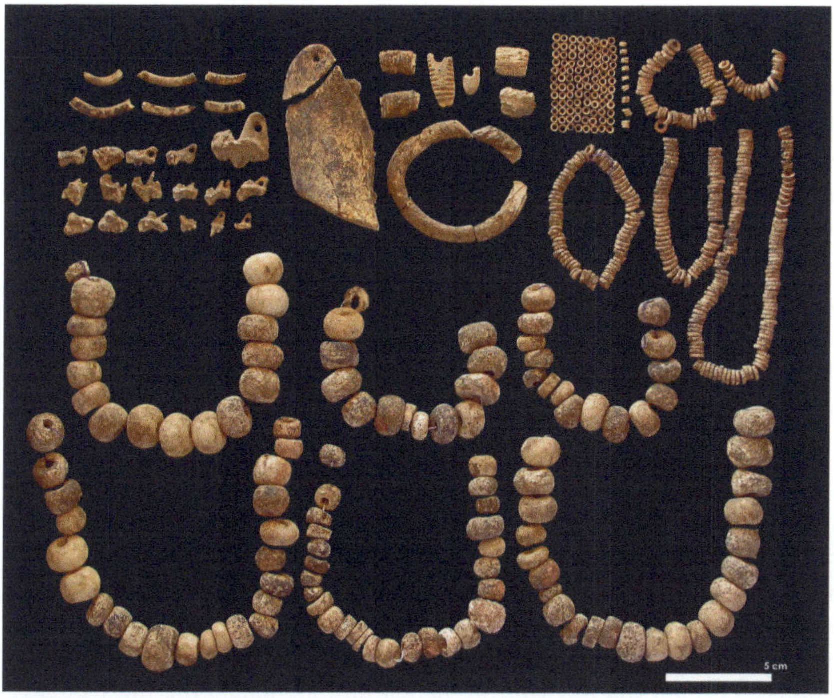

Fig. 8.6 Examples of the stone, ivory and bone adornments found with the person buried at Manemanya (Fig. 7, Sawchuk et al., 2019). *(Image by C. Klehm.)*

interspersed with the stone beads. At the mid-chest were large teeth from a lion and a hyena, with tooth roots perforated for use as a pendant. Near them was an oval flat stone, perforated at one end, with wavy lines incised perpendicular to its length. The left wrist bore a bracelet of 42 similarly perforated canid teeth (*Canis* sp., likely from jackals or wild dogs) and two narrow ivory bangles. The right arm bore a wide ivory bangle. Near the right shoulder lay a rather large item made of worked bone and ivory. The bone was perforated, so this item could have been worn or hung from the body. The size of the bone and ivory component is consistent with a large animal, like a hippo. Across the flexed legs and underlying one of them were lines of ostrich eggshell beads arranged in staggered dyads, like you might see on a leather placket with decorative beads sewn on in a geometric zig-zag pattern. All this was quite grand, but perhaps the most distinctive feature was the pelvic apron made of stone beads. Each of these perforated, rounded chunks of calcite had a real heft to it. Over 100 of these beads were strung around the pelvis. Although they had little color beyond some faint pink and brown tinges when we excavated them, they may have been far showier when freshly made and un-weathered. They would have been quite heavy. There were also large bone beads interspersed among the stone beads, but they were in such poor condition that their orientation couldn't be confidently assessed. The stone beads were found all around the body, so they were worn—not just laid on top. They did not spread out much beyond the torso, suggesting that the body may have been wrapped in a shroud.

The story

A rather tall young woman who had a noticeably asymmetrical face died of unknown causes. In the parlance of her society, she had access to goods of high value. Several of the goods related to dominance over the wild animals of the region: lion, hyena, jackal, hippo. There is no indication from her garb that her society valued the animals that we think they were herding: cattle, sheep, goats. The necklace of calabash-shaped stone beads could suggest some association with the world of healing, since small gourd containers are strongly associated with traditional medicine in Eastern Africa today. When she died, a place was prepared for her in a manner that included lighting a small fire. She was dressed elaborately, perhaps more elaborately than she would have been in life. Her body may have been wrapped, perhaps in an animal skin, and she was laid on the flat ground, where the fire had been. Then many, many strings of ostrich eggshell beads were strewn over her body. After that, large stones were placed onto the body, perhaps dropped with some force, based on the breaks seen on the skull. At some point thereafter, a massive pile of rocks was constructed around her. How her burial corresponds with those of other people whose skeletons are encased within the same monument remains to be determined.

The still incomplete picture

The fragmentary remains of the Manemanya person include the petrous portion of a temporal bone, the preferred site for palaeogenomic research. As I write this, samples of that tissue are being processed for aDNA. While aDNA is successfully extracted from fewer than half of all samples from tropical locales like Turkana, I hope that at least some bits of intrinsic DNA will be found. If the aDNA says that our sex determination was wrong, I will appreciate the correction.

We know from our brief excavations that there are other skeletons within the Manemanya structure. There wasn't time to explore the others. It is important to remember that this one person may or may not be unique within her society. We just don't know. Further, we don't know what kind of social structure or subsistence system generated this place. We can hazard some guesses—or construct some hypotheses, to put it more scientifically. The people living along the shores of Lake Turkana about 5–4000 years ago were doing quite well for themselves in many respects. The gifts of myriad strands of beads represent an impressive amount of labour and a form of material wealth (Coupland et al., 2016). Ostrich eggshell beads are known from much earlier African sites, so their presence is not unexpected. The people interred at the pillar sites also wore a diversity of stone ornamentation, which is both novel and distinctive. The community had the skills and the discretionary time to carve very hard, resistant materials into jewelry using differential resistance grinding methods. It seems unlikely that the teeth and ivory used as raw material for adornment were all "found objects." Rather, the community appears to have successfully hunted dangerous, large animals, then displayed prominently the evidence of that success. It seems improbable that a young adult would have personally downed the lion, the hippo, the hyena and other animals with which the body was adorned. The adornment, combined with the generous gifts, suggest that she was of high status. Perhaps that status had been earned during her brief lifetime, or perhaps was inherited. If it were the latter, that suggests that this was a community with clear differentiation of social roles and ranks.

There were enough people within a group—or a collection of related groups—that ideas of building monuments to encase and honor deceased group members could be planned, designed and executed. They constructed similar monuments on the east and west shores of Lake Turkana, at more or less the same time. This suggests that they had substantial watercraft, or they trekked long distances, or community connection was maintained via settlements interspersed all along the shore (for which there is negligible evidence). This was not a simple life, nor was it precarious. Here's what concerns me about what is known so far: if they had rather recently adopted domesticated animals and were relying on the products of those animals for food and clothing, why are those animals not the focus of the iconography? Was it like, "We have decided that pastoralism is the secure way to make a living, but we still want to see ourselves as hunters"? My co-workers have suggested that this was a time in which social and economic strategies were shifting around, and maybe constructing monuments helped to solidify things (Sawchuk, Goldstein, Grillo, & Hildebrand, 2018).

It is clear that many unanswered questions remain. This is the type of partial story that drives researchers to prioritize their time and energy, hoping to find the answers. You have to be more than a little bit driven to organize the money, the equipment, the campsites, the prophylactic medicines, the permits, and the tickets for flights to Lodwar. All over the world there are similar teams, exploring stories of past peoples whose lives are shrouded in mystery. Researchers' attention may be drawn to the artifacts, to the settings that people chose, or to the ways past groups organized themselves. The inclusion of osteobiography is an important part of those explorations. Study of the skeletal evidence provides a powerful reminder of the people themselves. Osteobiographies enhance our sense of the unique lives that contributed to humanity's journey.

References

Coupland, G., Bilton, D., Clark, T., Cybulski, J. S., Frederick, G., Holland, A., ... Williams, G. (2016). A wealth of beads: Evidence for material wealth-based inequality in the Salish Sea region, 4000–3500 CAL BP. *American Antiquity*, *81*(2), 294–315. https://doi.org/10.7183/0002-7316.81.2.294.

Grillo, K. M., & Hildebrand, E. A. (2013). The context of early megalithic architecture in Eastern Africa: The Turkana Basin c. 5000-4000 BP. *Azania-Archaeological Research in Africa*, *48*(2), 193–217. https://doi.org/10.1080/0067270x.2013.789188.

Hildebrand, E. A., & Grillo, K. M. (2012). Early herders and monumental sites in Eastern Africa: Dating and interpretation. *Antiquity*, *86*(332), 338–352. https://doi.org/10.1017/s0003598x00062803.Hildebrand et al., 2018Hildebrand, E. A., Grillo, K. M., Sawchuk, E. A., Pfeiffer, S., Conyers, L. B., Goldstein, S. T., ... Wang, H. (2018). A monumental cemetery built by Eastern Africa's first herders near Lake Turkana, Kenya. In *Proceedings of the National Academy of Sciences*. https://doi.org/10.1073/pnas.1721975115. 6 pp. + Supporting Information.

Lahr, M. M., Rivera, F., Power, R. K., Mounier, A., Copsey, B., Crivellaro, F., ... Foley, R. A. (2016). Inter-group violence among early Holocene hunter-gatherers of West Turkana, Kenya. *Nature*, *529*(7586), 394. https://doi.org/10.1038/nature16477.

Lundy, J. K., & Feldesman, M. R. (1989). The femur/stature ratio: A method for estimating living height from the femur. In *Paper presented at the 41st annual meeting of the American Academy of forensic sciences. Las Vegas, Nevada.*

Sawchuk, E. A., Goldstein, S. T., Grillo, K. M., & Hildebrand, E. A. (2018). Cemeteries on a moving frontier: Mortuary practices and the spread of pastoralism from the Sahara into Eastern Africa. *Journal of Anthropological Archaeology*, *51*, 187–205. https://doi.org/10.1016/j.jaa.2018.08.001.

Sawchuk, E. A., Pfeiffer, S., Klehm, C. E., Cameron, M. E., Hill, A. C., Janzen, A., ... Hildebrand, E. A. (2019). The bioarchaeology of mid-Holocene pastoralist cemeteries west of Lake Turkana, Kenya. *Archaeological and Anthropological Sciences*, *11*, 6221–6241. https://doi.org/10.1007/s12520-019-00914-4.

Wood, B., & Leakey, M. (2011). The Omo-Turkana Basin fossil hominins and their contribution to our understanding of human evolution in Africa. *Evolutionary Anthropology*, *20*(6), 264–292. https://doi.org/10.1002/evan.20335.

CHAPTER 9

Southern African foragers: Geographical and historical context

Prologue

The context of past lives represents a crucial aspect of osteobiography. In the case of Khoe-San hunter-gatherers, that context is multilayered (Fig. 9.1). Contemporary descendants represent a lineage that survived and increased in number for hundreds of millennia, given the genomic and palaeogenomic evidence (Chapter 7). They also represent a social group that has been characterized as marginal to the very lands over which their ancestors had mastery. If I was to study the bioarchaeology of the ancestral hunter-gatherers, I would also need to study and engage with contemporary issues.

I had decided that South Africa would be a place where I could study the ways that past hunter-gatherers survived—and even thrived—for millennia. This is an important element in the human evolution story. Without the creativity and innovation shown by past hunter-gatherer-foragers, the human lineage would have been just another dead end. This type of society is typical of about 95% of our history as a species (Binford, 2001). A defining feature of our lineage is the presence of simple stone tools that represent aids to food procurement. The date of the earliest stone tools in the world is debated, but it may be even older than the origin of *Homo*, which dates to about 2.8 million years ago (Chazan, 2018). We can use this as a conservative marker for how long humans have been using their brains and their tools to fill their stomachs. My focus is on the latter part—the very tail end—of this foraging continuum.

The problem for me, as I set out, was that I knew very little about South Africa, but that problem could be remedied. On my second visit to the country, in 1994, I had the chance to join the undergraduate field school offered by the Department of Archaeology at the University of Cape Town. Base camp would be a farmhouse near the little market town

Osteobiographies
https://doi.org/10.1016/B978-0-12-823880-6.00010-0

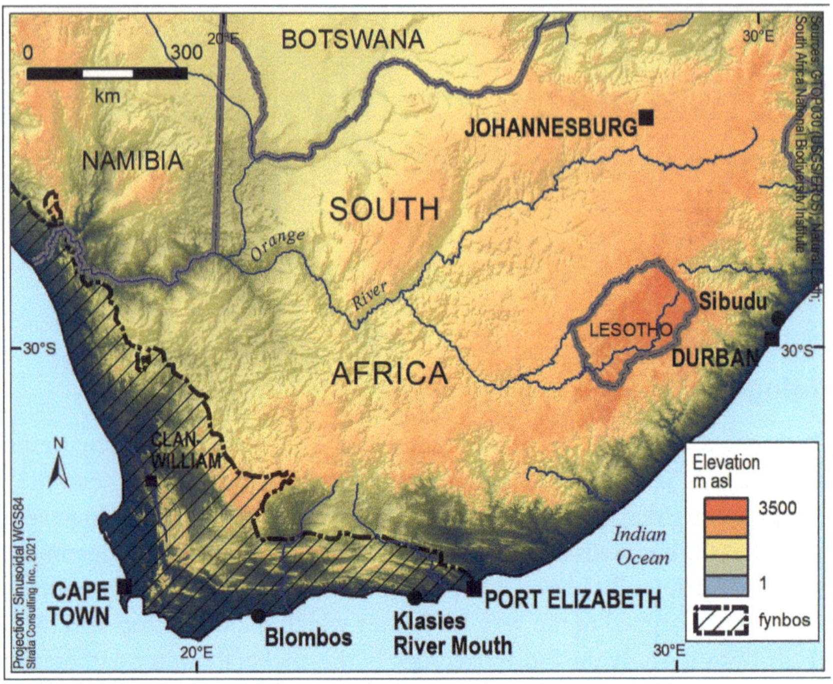

Fig. 9.1 Map of the region discussed in this and subsequent chapters.

of Clanwilliam, about 230 km (140 mi) north of Cape Town. We would explore the rock shelters of the Cape Fold Mountains in the sector known as the Cederberg, along the Olifants River. In this region, the only human inhabitants prior to Europeans were ancestors of today's Khoe-San people (Box 9.1). They had been living there for at least 10,000 years and probably much longer. This was back when elephants, eland, kudu and a host of other animals roamed the valley. The structure of this archaeology field school was unique. Field schools normally train students in excavation methods and artifact typology. Students learn about what kind of tools people made, what foods and raw materials would have been available. In the Cederberg, excavation and subsistence were not the initial topics. Instead, the major focus was on rock art. The walls of many of the natural rock shelters, eroded out of the ancient sandstone, had been used by past hunter-gatherers to conceptualize their views of the world. On those walls there were not only beautiful game animals, but also images of human and partially human figures captured in dynamic poses that have a spiritual dimension. This was a field school experience that opened the

BOX 9.1 Contemporary San communities and names

While the focus of this chapter is on the ways that archaeological scholarship can enrich our understanding of the past, there are strands of more recent history and modern life that are important, too. There are contemporary San communities in Angola, Botswana, Lesotho, Eswatini, Namibia, South Africa, Zimbabwe, and Zambia, numbering in total about 130,000 (Hitchcock, 2020). Some pursue foraging lifeways, at least to some extent. They are collectively represented by the Working Group of Indigenous Minorities in Southern Africa (WIMSA), the South African San Council (SASC), and other regional groups. The primary concerns of these organizations are issues of land and resource rights. Charitable organizations like the Kalahari Peoples Fund (KPF) and Cultural Survival also support community-building and engage in lobbying on their behalf.

Globally, public recognition of the San grew during the 20th century, thanks to stories about the Bushmen of the Kalahari. Books and other public initiatives by Sir Laurens van der Post (van der Post, 1961), the Marshall family (Marshall, 1958), Marjorie Shostak (Shostak, 1981), the Harvard Kalahari project (Archaeology Centre, 2014; Lee, 1979) and other scholars (e.g., Silberbauer, 1981), plus *The Gods Must be Crazy* (film 1980, sequel 1989) made the "Bushman" an icon of the African hunter-gatherer. Within the academic world, ethnographic fieldwork in the Kalahari region formed the basis for an influential set of ideas subsumed under the label "man the hunter" (Lee & DeVore, 1968), and subsequently a challenge to the validity of the underlying premises, dubbed "the great Kalahari debate" (Solway & Lee, 1990). Meanwhile, the San were coping with hostile governments, dangerous military service, restrictions on group movements and hunting, and a host of other challenges. The Khoi, or Khoekhoe, emerged from San groups, between 2000 and 1200 years ago, through mixing with incoming migrant herders from East Africa (see Coutinho et al., 2021 for a summary). In apartheid South Africa, the San and the Khoekhoe were classified as "coloreds" and were offered only rudimentary education, solely in Afrikaans. There was little encouragement to maintain cultural identities or languages.

In Botswana, they are called Basarwa, based on a Bantu language term for "those who do not rear cattle." The term Khoisan, as a collective term for the pastoralist Khoi and the foraging San, entered common parlance through the writings of anthropologist Isaac Schapera (Schapera, 1930). This is still the term for their distinctive "click" languages, when considered collectively. The word Khoi or Khoe (a newer linguistic form) means "person" in Nama. The word San is the Khoe word for "foragers" or "bushmen" (Barnard, 1992).

The establishment of the new South African constitution and bill of rights in 1996 has encouraged many persons to claim and take pride in their indigenous heritage. In 2002, at a meeting of WIMSA and SASC, the San expressed their wish to be referred to by their individual community names (see Fig. B9.1.1) or collectively as San. When collectively referring to the San and the Khoe, the term

Continued

BOX 9.1 Contemporary San communities and names—cont'd

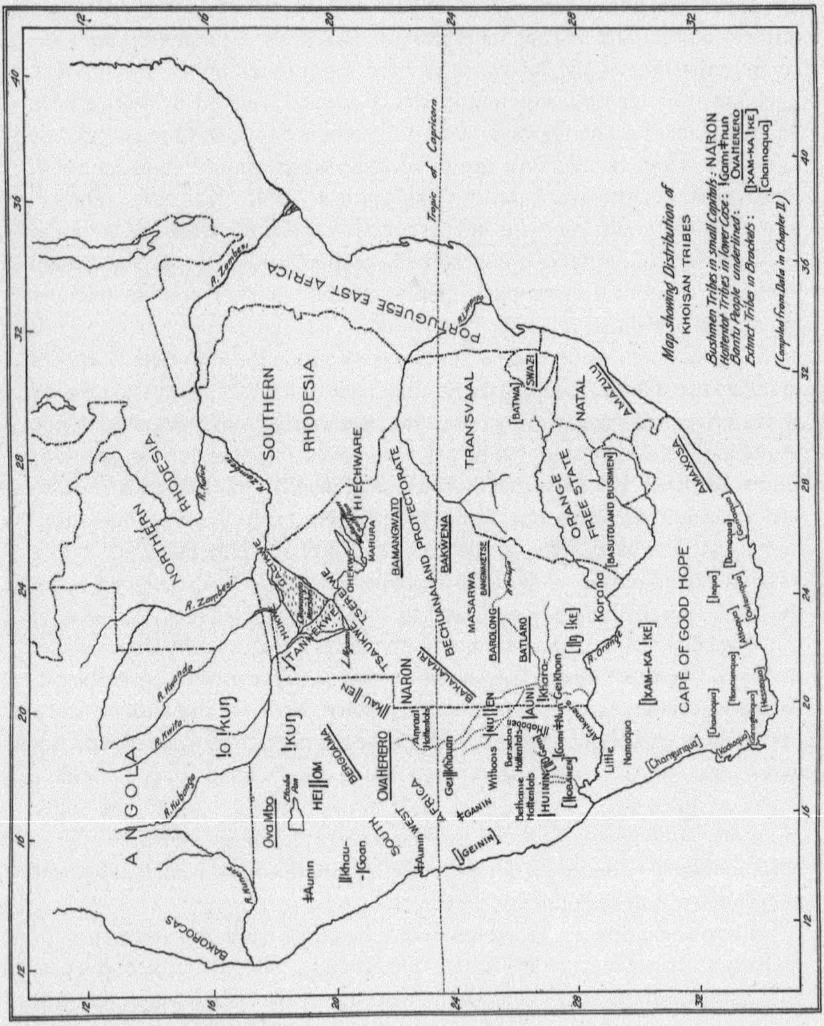

Fig. B9.1.1 Map showing the distribution of "Khoisan Tribes" (Schapera, 1930). By the time this map was drawn, most groups plotted along the southern coast were categorized as extinct.

BOX 9.1 Contemporary San communities and names—cont'd

Khoe-San was suggested (Crawhall, 2006). Genomic and linguistic evidence for connections between the Khoe and San, of ancient but undetermined time-of-origin, provide the rationale for referring to their ancestors collectively, as ancestral Khoe-San (or KhoeSan).

Archaeology Centre, U. o. T. (2014). *Kalahari Reflections*. https://www.youtube.com/watch?v=EgFhtlwbX0Y.

Barnard, A. (1992). Ethnic classification, origins, and history of the Khoisan peoples. In *Hunters and Herders of Southern Africa* (pp. 16–38). Cambridge: Cambridge Studies in Social and Cultural Anthropology.

Coutinho, A., Malmstrom, H., Edlund, H., Henshilwood, C. S., van Niekerk, K. L., Lombard, M., … Jakobsson, M. (2021). Later Stone Age human hair from Vaalkrans Shelter, Cape Floristic Region of South Africa, reveals genetic affinity to Khoe groups. *American Journal of Physical Anthropology*. https://doi.org/10.1002/ajpa.24236.

Crawhall, N. (2006). Languages, genetics and archaeology: Problems and the possibilities in Africa. In H. Soodyall (Ed.), *The Prehistory of Africa* (pp. 109–124). Johannesburg: Jonathan Ball Publishers.

Hitchcock, R. K. (2020). The plight of the Kalahari San: Hunter-gatherers in a globalized world. *Journal of Anthropological Research Summer, 2020*, 164–184.

Lee, R. B. (1979). *The !Kung San: Men, Women and Work in a Foraging Society*. Cambridge: Cambridge University Press.

Lee, R. B., & DeVore, I. (Eds.). (1968). *Man the Hunter*. Chicago: Aldine Publishing Company.

Marshall, E. M. (1958). *The Harmless People*. New York: Knopf.

Schapera, I. (1930). *The Khoisan peoples of South Africa*. London: Routledge and Kegan Paul Ltd.

Shostak, M. (1981). *Nisa: The life and words of a !Kung woman*. Cambridge: Harvard University Press.

Silberbauer, G. (1981). *Hunter and habitat in the Central Kalahari Desert*. Cambridge: Cambridge University Press.

Solway, J. S., & Lee, R., B. (1990). Foragers, genuine or spurious? *Current Anthropology, 31*(2), 109–146.

van der Post, L. (1961). *The Heart of the Hunter*. Hammondsworth: Penguin.

learners' minds to the esoteric, spiritual side of past peoples' lives and to the beliefs that guided their lives.

The town of Clanwilliam has a main street of Cape Dutch buildings, trailing downhill from the Dutch Reform Church. Afrikaans is the language of commerce. Like all South African towns of that era, there were sections for the whites and for the "coloreds." The town provided goods and services for the workers from the region's citrus and rooibos tea farms, and it hosted tourists coming to see the plants of the Cederberg Conservancy. The Conservancy and the broader area are part of a biodiversity hotspot: the fynbos biome, also called the fynbos biotic province or Cape Floristic Region. One fynbos plant, rooibos tea (*Aspalathus linearis*), is now globally popular, to Clanwilliam's benefit. In 1994, rooibos was a mostly local product and Clanwilliam was a pretty sleepy place. The juxtaposition of the colonial town and the rock art illustrated neatly how the descendants of the people who had lived in the Cederberg region before the arrival of Europeans had become the workers on the tea farms and in the citrus groves, living in shacks on the edges of the town.

One evening, we keen archeologists set off to see a rock shelter. The novelty was that we would view it in limited light, as the creators of the images likely would have wished. Some delicate painted connections among the images are easier to see in glancing light. We would use battery-powered light to mimic firelight. As we scrambled across the loose boulders of the Cederberg in the darkness, my foot slipped, and my leg dropped into a crevasse. I righted myself and moved on, but I had gouged my lower leg rather dramatically. After our outing, back at the farmhouse, it was decided that my leg needed stitches. An archeologist would drive me to town. At around midnight, he rang the bell at the clinic. A nursing sister quickly stitched up my wound. I hoped we could then return to the farmhouse, but we were expected to wait for the physician. In due course, a sleepy-looking white man wearing a golfing jacket arrived. My archeologist host explained that I was a visitor from Canada, interested in the past peoples of the region. At this, in an annoyed tone, the physician boomed, "You are not one of those people who claim that THEY were here FIRST?!"

Narratives about "who was here first" and assertions that colonists "settled on unoccupied land" are common in many parts of the world. I was familiar with this rationale for settler take-overs from my North American home. Nevertheless, I was surprised to encounter it in Southern

Africa. Five hundred years ago, Arab traders on the Indian Ocean side and Dutch and English colonists on the Atlantic Ocean side encountered and displaced people who had occupied the landscape for many thousands of years. The way this displacement and subjugation was shaped into apartheid-era history affected everything, even archaeology. The scar on my lower leg is faint now, but I retain a vivid memory of this lesson in South African history.

The remarkable shrinking forager

The complex history of South Africa has been shaped by human ambitions and global events, but also by landscape. My focus is on the ancestral Khoe-San people who may have lived across much of the continent in the very distant past. Their society was one of hunting-and-gathering (I will used "foraging" as a short form), which by anthropological definition means they were residentially mobile groups of small numbers, which were egalitarian in social organization. They subsisted on naturally occurring resources. By about 10,000 years ago Khoe-San foragers had apparently found great success in the fynbos biome. This region (roughly 80,000 km^2) of unique plants is the basis of an ecosystem which is mostly within the Western Cape Province of South Africa. The boundaries of the fynbos biome begin at the coastline, and extend landward, including coastal forelands and the Cape Fold Mountains, or Cape Fold Belt (Allsopp, Colville, & Verboom, 2014). This overview will focus on that part of South Africa, placed into its broader geographic context.

The archeological story

International tourists to South Africa often combine their viewing of the fynbos protea and erica blooms around Cape Town with a visit to the Cradle of Humankind, just outside Johannesburg. That is the region that has provided evidence of ancestors preceding modern *Homo sapiens*, from early discoveries of Australopithecines to recent discoveries of proposed lineages like *Homo naledi* (Wood, 2019). The era during which the earliest stone artifacts were produced is called the Early (or Earliest) Stone Age (ESA), or the era of mode 1 technology (Clarke, 1969). The earliest stone tools found in South Africa date to almost 2 million years ago (Chazan et al., 2012). Typical ESA tools, like "hand axes" have been discovered across much of the landscape. The genomic divergence of the Khoe-San lineage has been estimated at 350–260,000 years ago

(Fan et al., 2019; Schlebusch et al., 2017). Scant fossil evidence has been found in South Africa for the transition from earlier forms of *Homo* to anatomically modern *Homo sapiens*, which we think occurred about 200–300,000 years ago. Middle Stone Age (MSA) technology from this time period suggests similar patterns of behaviors across diverse human populations (Bergström, Stringer, Hajdinjak, Scerri, & Skoglund, 2021). There are wonderfully productive South African sites that document the lives of the MSA moderns from a bit more than 100,000 years ago (Lombard et al., 2012; Mitchell, 2002).

In strong alignment with genomic expectations, some discoveries at Middle Stone Age rock shelters like Klasies River Mouth, Blombos, Pinnacle Point and Sibudu Rock Shelter show some similarities with cultural practices of Khoe-San communities in the modern era. The discovery at Blombos of abalone shells holding pigments and grinders suitable for creating rock art (Henshilwood et al., 2011) and the discovery at Sibudu of arrow points (Backwell et al., 2018) are examples of possible technological continuity. This is not to say that there are not differences between ancestral groups and the people who provide historic/modern examples of these artifacts in use (d'Errico et al., 2012; Pargeter, MacKay, Mitchell, Shea, & Stewart, 2014). Rather, it signals the sustained use of some items as humans shifted in their locales and approaches to survival. The ways in which objects were used, and the meanings given to objects and symbols may have changed through the millennia. Nevertheless, some form of continuity is apparent. In parallel with the material record, some of the fragmentary human remains from Klasies River Mouth appear similar to those of more recent Khoe-San skeletons (Cameron, Pfeiffer, & Stock, 2021; Rightmire & Deacon, 1991), although here, too, more information from future discoveries may shift our perspective.

How might fragmentary human remains be seen as belonging to a particular group of people? Here is one key part of the Khoe-San story: they are small (Fig. 9.2). Their average adult stature and body mass, their rounded crania and relatively delicate facial structure, and even their teeth are toward the petite end of the human distribution (Black, Ackermann, & Sealy, 2009; Pfeiffer, 2012). These distinctive features can be tracked back about 9000 years through study of radiocarbon dated skeletons, but it is not clear how much further back in time they go. Morphology (size and shape) can and does change with time.

Fig. 9.2 Image of the director of the Denver African Expedition of 1925, with "two average-sized Bushmen." (Gordon, 1997). (Image from Denver Municipal Facts, January–February 1927, Vol. X, Nos. 1 and 2, page 13. Photo by P. Hoefler.)

At around 40,000 years ago, the technological package shifted to the Later Stone Age (LSA) (Box 9.2). The Later Stone Age is thus the most recent end of one of the longest and best records of hunter-gather lifeways anywhere in the world. At a broad-brush level, the stone tool categories remain much the same, but there is a greater presence of items made from bone and shell. As in the MSA, different times and places show characteristic assemblages. The presence of points and link shafts in the LSA toolkit likely represent hunters' reliance on neurotoxic mixtures made and stored with great care, delivered to prey by light-draw bows. Using the strategy of persistence hunting, even large animals could be killed. Pottery is not a part of the toolkit until after 2000 years ago. As in the MSA, many of the most productive archeological sites are found along the coast. There are time gaps in both the MSA and LSA records because coastlines were rising and falling. By about 10,000 years ago people who looked

BOX 9.2 Later Stone Age material culture

Information about the tool kit of Later Stone Age (LSA) peoples of southern Africa remains incomplete, as is so often the case with cultures of the distant past. Nevertheless, we do know quite a bit (Mitchell, 2002), thanks to four complementary lines of evidence. The focus for this brief summary will be the latter part of the Holocene—the past 6000 years—prior to the dramatic disruptions associated with the arrival of Europeans.

- There is apparent continuity of some cultural features—like use of ochre and arrows suitable for poison delivery—for thousands of years so we can cautiously compile the evidence over several millennia (d'Errico et al., 2012; but see also Pargeter et al., 2014).
- Occupational sites are sometimes well preserved, especially in the rock shelters of the Cape Fold Belt. They provide evidence of bedding, food waste and tool making—although the intensity of the shelters' utilization is often unclear. Shell middens along the coast help expand evidence of daily activities (Mackay, 2016; Parkington, 2006).
- Because several key artifact types are very similar to historically documented Khoe-San tool kits, ethnographic records can be useful.
- Finally, and most distinctively, panels of rock art found throughout the Cape Fold Belt provide not only information about garb and tools, but also a kind of window into the mindset and belief system that provided structure to people's lives and guided their decisions (Deacon & Deacon, 1999; LewisWilliams, 2015; Parkington & Dlamini, 2015) (Fig. B9.2.1).

Daily tasks appear to have focused on food procurement. Mobility is generally not compatible with food storage and there were no managed crops. Detailed knowledge of the landscape and climate formed the basis for systematic, likely cyclical, exploitation of reliable resources. Rock art, patterns of bone strength, and ethnography strongly suggest that hunting was a man's task, and foraging/gathering was a woman's task. Important raw materials included skins and sinew for clothing and bags; stone suitable for manufacturing diverse kinds of useful objects, like points, drills and scrapers; other types of stone to be ground into variably-sized, doughnut-shaped digging stick weights; ostrich eggshells that were emptied and used whole as flasks or manufactured into small beads for stringing; hematite (ochre) and charcoal for pigment; bone from animals and birds for points, link-shafts and other tools and ornaments; tortoise shells for containers and bowls; and a wide array of plant materials for bows, arrow shafts, digging sticks, bedding and medicinal applications.

While spears were used for hunting in some regions, the light-draw bow and link-shaft arrow was ubiquitous. A link-shaft arrow is made of a light reed, a torpedo-shaped "link" made from bone or wood, a short reed collar and the point itself, to which poison is applied (Fig. B9.2.2). Hunting success using this weapon is contingent on relatively close access to prey, and delivery of a neurotoxic poison where the point penetrates the animal's body. Great care was required for the manufacture from plant and animal toxins, the storage and application

BOX 9.2 Later Stone Age material culture—cont'd

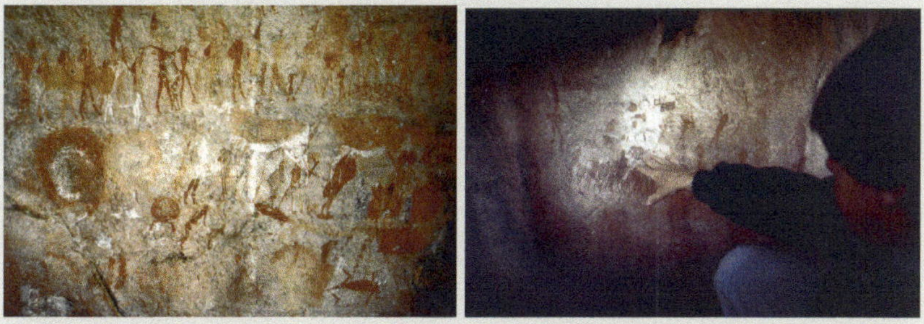

Fig. B9.2.1 (A, B) Rock art panels created by Later Stone Age people; these are in the Cederberg region of the Western Cape, South Africa. The images of people, animals, and beings with unique features provide unique and sometimes puzzling clues about their lives and their beliefs. (Photos: S. Pfeiffer.)

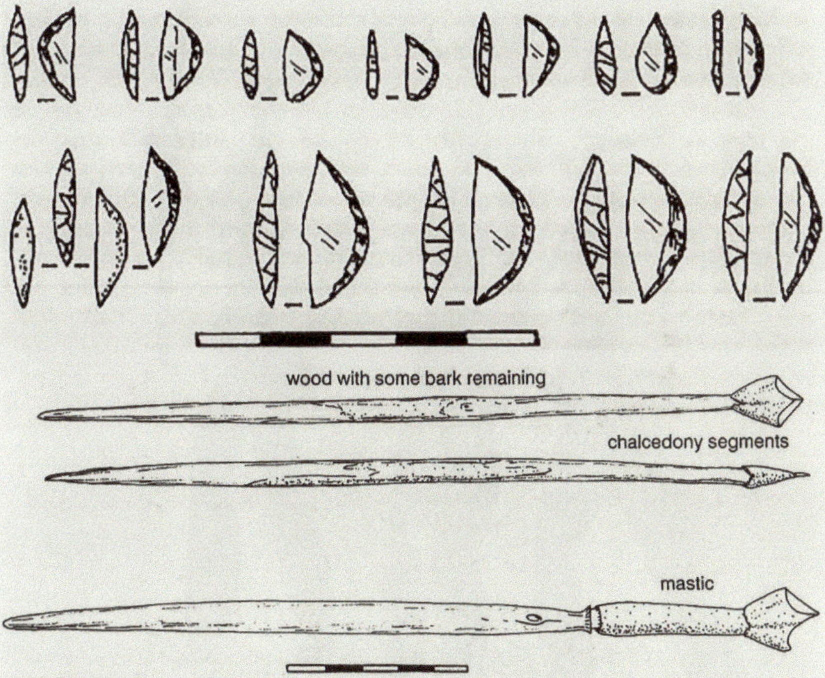

Fig. B9.2.2 (A) Examples of stone segments from archeological contexts. The scale = 50 mm. (B) Two of the demonstration arrows made for Bleek and Lloyd in 1875, now in the Pitt Rivers Museum, Oxford. The scale = 50 mm. ((A, B) Both images reproduced with permission from Deacon, J. (1992). *Arrows as Agents of Belief Among the /Xam Bushmen.* Cape Town: South African Museum.)

Continued

BOX 9.2 Later Stone Age material culture—cont'd

of the poison. It was so dangerous that it was not applied to arrow tips, since a hunter could nick himself. After the poison entered its blood stream, the prey was tracked until the poison disabled the animal. The manufacture, the careful storage and application of this substance was a central feature of a hunter's life (Deacon, 1992). Through archaeology, this technology can be traced back over 60,000 years (Backwell et al., 2018). The ubiquitous womens' tools were containers of various types and digging sticks. Especially when supplemented with a weight to enhance its forward momentum, a digging stick could be used to dig nutritious corms and roots, to bash limpets off seaside rocks, to enlarge openings to animal burrows, and so on. Like arrow points, digging stick weights have substantial antiquity (Ouzman, 1997) (Fig. B9.2.3).

Fig. B9.2.3 Tracing of a rock painting of three women carrying digging sticks weighted with bored stones. Black represents red. Scale bar is 30 mm. (Credit: Ouzman 1997; reproduced with permission of National Museum, Bloemfontein.)

BOX 9.2 Later Stone Age material culture—cont'd

Backwell, L., Bradfield, J., Carlson, K. J., Jashashvili, T., Wadley, L., & d'Errico, F. (2018). The antiquity of bow-and-arrow technology: evidence from Middle Stone Age layers at Sibudu Cave. *Antiquity, 92*(362), 289–303. https://doi.org/10.15184/aqy.2018.11

d'Errico, F., Backwell, L., Villa, P., Degano, I., Lucejko, J. J., Bamford, M. K., … Beaumont, P. B. (2012). Early evidence of San material culture represented by organic artifacts from Border Cave, South Africa. *Proceedings of the National Academy of Sciences of the United States of America, 109*(33), 13,214–13,219. https://doi.org/10.1073/pnas.1204213109

Deacon, H. J., & Deacon, J. (1999). *Human beginnings in South Africa.* Cape Town: David Philip Publishers.

Deacon, J. (1992). *Arrows as Agents of Belief Among the /Xam Bushmen.* Cape Town: South African Museum.

LewisWilliams, J. D. (2015). *Myth and Meaning: San-Bushman Folklore in Global Context.* Routledge.

Mackay, A. (2016). Three arcs: observations of the Elands Bay and northern Cederberg landscapes. *Southern African Humanities, 29,* 1–15.

Mitchell, P. (2002). *The Archaeology of Southern Africa.* Cambridge, U.K.: Cambridge University Press.

Ouzman, S. (1997). Between margin and centre. In L. Wadley (Ed.), *Our gendered past: Archeological studies of gender in Southern Africa* (pp. 71–106). Johannesburg: Witwatersrand University Press.

Pargeter, J., MacKay, A., Mitchell, P., Shea, J., & Stewart, B. A. (2014). Primordialism and the "Pleistocene San" of southern Africa. *Antiquity*, 1072–1079. https://doi.org/10.15184/aqy.2016.100

Parkington, J. E. (2006). *Shorelines, Strandlopers and Shell Middens.* Cape Town: Creda Communications.

Parkington, J. E., & Dlamini, N. (2015). *First People* (Vol. 5). Cape Town: Krakadouw Trust.

and behaved like ancestral Khoe-San people were living in rather large numbers throughout the fynbos region and beyond, through the interior and extending east to that coastline. Then, beginning at around 2000 BCE, several changes occurred.

Along the west side of South Africa, a few sites after this time have provided not only potsherds but also bones of caprines—domesticated sheep and/or goats (Lander & Russell, 2018). Combined with the genomic reconstructions, these clues represent the arrival of Khoikhoi herders, who were related to the San foragers but appear to have interacted with more northerly groups to acquire a new subsistence strategy (Sadr, 2013). It is not clear how the Khoikhoi and the foragers interacted. By the time the Dutch explorer Jan Van Riebeeck reached the Cape of Good Hope in 1652, the Khoikhoi were also herding cattle. Thereafter, they supplied seafarers and developed a commercial relationship with the newcomers (Boonzaier, Malherbe, Berens, & Smith, 1996; Elphick, 1977).

About 1700 years ago, outside the fynbos region on the east side of the country, the migration of Bantu-speaking farmers reached and crossed the Shashi and Limpopo Rivers (de Maret, 2013). This was an expansion that had begun in the Cameroon-Nigeria borderlands about 4000 years ago. Archeologically, this is characterized as the Iron Age (IA), subdivided by dates and ceramic traditions (Mitchell, 2002). Cultivating sorghum, millet, cowpeas and other vegetables (later adding maize), raising cattle and acquiring the skills to smelt iron, the Bantu-speaking peoples represent a large number of distinct communities. Those who ventured south include ancestors of the Venda, Xhosa, Sotho, Tswana, Tsonga, Swati, Ndebele, and Zulu, all of whose languages are today among the official languages of South Africa. Collectively, they challenged the foragers for access to good grazing lands and pushed the foragers into the Drakensberg Mountains of what is now Lesotho and KwaZulu-Natal. Genomic studies of contemporary descendants document sometimes substantial Khoe-San admixture, reflecting the prolonged nature of those encounters (Schlebusch, Prins, Lombard, Jakobsson, & Soodyall, 2016). Over a millennium later, Europeans encountered well organized land-holding farmers with iron weapons who mounted stiff resistance to their encroachment. Perhaps not surprisingly, in the eastern half of the country the colonialists treated the Bantu-speaking groups as the Indigenes. The prior claims of the Khoe-San peoples were forgotten or ignored.

The Europeans make their presence felt

Meanwhile, on the western side of the country, the Khoe-San peoples—both San and Khoikhoi—met the Europeans. While Europeans developed a commercial relationship with the Khoikhoi, their approach to the San was one of disdain and distrust. Soon a program of eradication was mounted. Poison arrows were no match for European weaponry, and the San were characterized as thieves. The famous philologist team who documented the languages and stories of the San—Wilhelm Bleek, his sister-in-law Lucy Lloyd, and his daughter Dorothea Bleek—found their subjects in the prisons of Cape Town and Robben Island (Barnard, 2007:23–37; Lewis-Williams,

2000). Missionaries sought to "civilize" the San; it was said that if captured at a young age they made good servants (Penn, 1996). The campaign of eradication was countrywide. The last "Bushman" in the Drakensburg Mountains was shot at the turn of the nineteenth century.

At the same time as a human lineage was being eradicated, there was a sense that the disdained aboriginal people of the region (labeled variously Caffir, Bushman, Hottentot, Strandloper) represented something biologically special. Both prior to Darwin's *Origin of Species* (Darwin, 1859) and subsequently, European and American scholars categorized the group as inherently distant from fully-realized human intelligence. This assumption applied to other Africans as well (Gould, 1981; Wolpoff & Caspari, 1997). However, they surmised that Khoe-San peoples were distinct from other Africans. In this context, the display of Sara Baartman, the famous "Hottentot Venus" was a great success in Europe. She remained a source of fascination after her death, as she was autopsied by the famous anatomist Cuvier. Sara Baartman (1789–1815, buried in South Africa in 2002 (Tobias, 2002)), was far from the only living person exported for display to the curious (see Chapter 12). There was also a vigorous interest in collection of skeletons—especially crania—for scholarly study (Bonner, Esterhuysen, & Jenkins, 2007; Legassick & Rassool, 2000). The discussion of how to classify these people shifted with changing perspectives. Well into the 20th century, their distinctive size and shape were said to justify a special racial category. Prominent anthropologist Carleton Coon proposed that humans evolved into five separate branches: Caucasoid, Congoid, Mongoloid, Australoid and Capoid (Coon, 1965) (see Chapter 7). While such categorizations were subsequently eclipsed by more biologically informed understandings of human variability, there remains no argument that Khoe-San peoples are distinctive, as reflected by their genomic uniqueness.

How did prehistorians of that era integrate this distinctiveness into narratives about the African past? An often-repeated origin story was that Khoe-San peoples were vestigial, that their petite frames and small faces represented the cumulative result of isolation and starvation. They were thought to be depauperate descendants of big people who shrunk, and it was further surmised that the shrinkage had happened rather recently. The name for the putative ancestral race was Boskop. The name comes from a large partial skull discovered in 1913. Anatomist Raymond Dart posited a large bodied, robust ancestral population of the "Boskop type" (Dart, 1923; Galloway, 1937). The size of a skeleton was taken to be an indication of its antiquity (FitzSimons, 1926). This caused considerable confusion, as archeological sites from ancestors of Bantu-speaking groups could be mistakenly classified as ancestral Khoe-San (Tobias, 1959) because the skeletons were deemed to be "Boskop." This muddled the picture considerably. The idea of a Boskop type quickly proved to be a scientific nonstarter (Rightmire, 1978; Singer, 1958). My colleague at University of Cape Town, Alan G. Morris, has vividly described the ideas and implications of this long colonialist-racist era (Morris, 2005, 2012, 2014). As a relative newcomer to the scene, I witnessed a gradual loss of interest in the quest for the large-yet-bushman skeletons, which continued into the late 20th century.

A large number of small people

A large number of human skeletons had been accessioned into the museums and university repositories of South Africa by the early 1990's. Some were the result of systematic excavations in rock shelters of the Cape Fold Mountains that not only sheltered burials but also provided occupational evidence of foragers lives. Many others had been discovered by chance in the coastal forelands, often in beach sand along the coast. The single, primary burials were of people of all ages. Those from rock shelters often included artifacts like painted stones or garb, like ostrich eggshell beads. Most of those from coastal forelands were not accompanied by artifacts or garb that would date or identify them culturally. Body sizes and their cranial shapes were consistent with expectations for Khoe-San ancestors. Early in my visits, I met a young scholar at the University of Cape Town, Judith Sealy, who was developing and applying methods of stable isotope analysis—quite new at that time—to assess their diets. She had organized radiocarbon dating of many of the skeletons, through the national laboratory in Pretoria. She certainly could have rebuffed the overtures of a keen Canadian who didn't know anything about South African archaeology. To my great good fortune, she did not. We were both interested in creating osteobiographies for the foragers who had survived and thrived in the fynbos region and beyond. Similar to my Canadian research of Huron-Wendat ancestors, there are many ethnographic and historic documents that describe Khoe-San people of the modern era. While not from the same times and places as the LSA skeletons, the history and ethnographies would help us construct testable hypotheses about ancestral Khoe-San lifeways.

Another pillar of research support was a catalogue of the many Holocene (that is, the past 10,000 years) human skeletons curated by South African institutions, as compiled by Morris (1992). Of course, the catalogue includes skeletons that are not ancestral Khoe-San. There were Iron Age farmers from those regions where they had become established. Narrowed down to discoveries in fynbos regions, several hundred likely Khoe-San skeletons were listed. They were held at a minimum of nine institutions, spread across the country.

They had been found—mostly as single, unadorned burials—across a huge landscape. Can they be considered collectively, as a single biological population? Following extensive research, the conclusion is yes, they can. That confident statement is built on the work of University of Cape Town colleagues who took on the challenging work of assessing cranial and dental morphology using modern tools of measurement and analysis. Deano Stynder used geometric morphometric methods to study the crania of hundreds of ancestral Khoe-San crania from archeological sites across South Africa. He concluded that there is impressive consistency of size and shape, both across space and through time (Stynder, Ackermann, & Sealy, 2007a, 2007b). Soon thereafter, Wendy Black completed an ambitious study of dental size and shape, comparing ancestral Khoe-San dentitions to those from other African communities. She, too, found internal consistency within a very large sample and also distinctiveness compared to other Africans (Black et al., 2009). To the extent that the size and shape of skulls and teeth reflect inheritance, we can rely on these studies to identify a distinctive, coherent biological population. Paleogenomics research has since confirmed the links between Holocene skeletons and Khoe-San descendants (Morris, Heinze, Chan, Smith, & Hayes, 2014; Schlebusch et al., 2017; Skoglund et al., 2017). Behaviorally, there were regional differences in matters like diet and toolkits. Biologically, it appears that there had not been significant or sustained barriers to the exchange of mating partners.

Bioarchaeology is the study of bones and other biological materials found in archeological contexts, done to provide information about human lives in the past. As such, study of the ancestral Khoe-San skeletons would focus on their world long ago. However, this work cannot be divorced from the historical acts and ideas that still affect how their descendants are regarded today. Those acts and ideas also affect how descendants view their own heritage. Our work has refuted three fallacies of the historic era: that the land was "empty" when the newcomers arrived, that small adult body size is a side-effect of malnutrition or degeneration, and that Khoe-San foragers lived in unsuccessful communities. The application of current scientific methods to large numbers of skeletons (Fig. 9.3) has built a clear

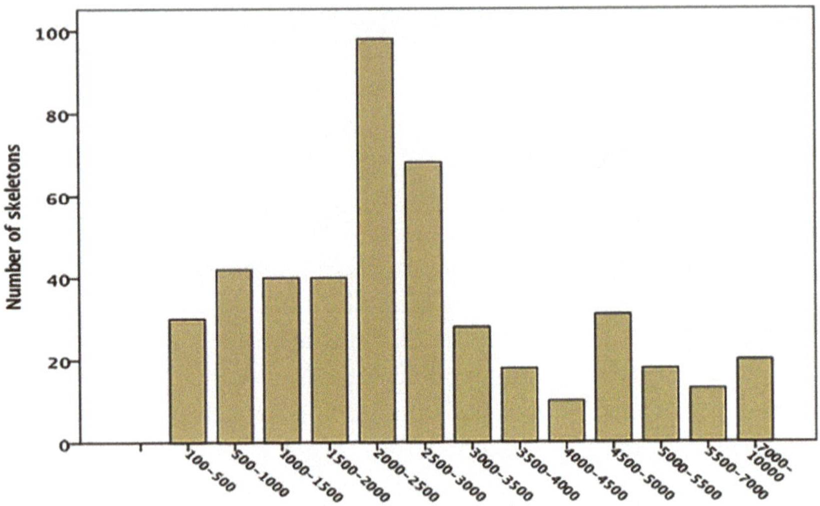

Fig. 9.3 A plot of radiocarbon dates from 459 ancestral Khoe-San skeletons, organized into bins of time since the present. The rarity of skeletons from more than 5000 years ago is probably linked to higher sea levels that washed away evidence of human activity. The decline in numbers between 2000 and 1000 years ago remains unexplained. It may be linked in some way to the arrival of new people.

picture of a population that was well fed, increasing in number, and adaptively organized, although not without internal conflicts.

References

Allsopp, N., Colville, J. F., & Verboom, G. A. (Eds.). (2014). *Fynbos: Ecology, evolution, and conservation of a megadiverse region*. Oxford, UK: Oxford University Press.

Backwell, L., Bradfield, J., Carlson, K. J., Jashashvili, T., Wadley, L., & d'Errico, F. (2018). The antiquity of bow-and-arrow technology: Evidence from middle stone age layers at Sibudu Cave. *Antiquity*, *92*(362), 289–303. https://doi.org/10.15184/aqy.2018.11.

Barnard, A. (2007). *Anthropology and the Bushman*. Oxford: Berg.

Bergström, A., Stringer, C., Hajdinjak, M., Scerri, E. M. L., & Skoglund, P. (2021). Origins of modern human ancestry. *Nature (London)*, *590*(7845), 229–237. https://doi.org/10.1038/s41586-021-03244-5.

Binford, L. R. (2001). *Constructing frames of reference: An analytical method for archaeological theory building using ethnographic and environmental data sets*. Berkeley: University of California Press.

Black, W., Ackermann, R. R., & Sealy, J. C. (2009). Variation in Holocene Khoesan dentition. *American Journal of Physical Anthropology*, *S48*(138), 92–93.

Bonner, P., Esterhuysen, A., & Jenkins, T. (Eds.). (2007). *A search for origins: Science, history and South Africa's 'cradle of humankind'*. Johannesburg: Wits University Press.

Boonzaier, E., Malherbe, C., Berens, P., & Smith, A. (1996). *The cape herders: A history of the Khoikhoi of Southern Africa*. Cape Town & Johannesburg: David Philip.

Cameron, M. E., Pfeiffer, S., & Stock, J. T. (2021). Small body size phenotypes among Middle and Later Stone Age Southern Africans. *Journal of Human Evolution*, *152*, 102943. https://doi.org/10.1016/j.jhevol.2020.102943.

Chazan, M. (2018). *World prehistory and archaeology: Pathways through time* (4th ed.). New York: Routledge, Taylor & Francis Group.
Chazan, M., Avery, D. M., Bamford, M. K., Berna, F., Brink, J., Fernandez-Jalvo, Y., … Horwitz, L. K. (2012). The Oldowan horizon in Wonderwerk Cave (South Africa): Archaeological, geological, paleontological and paleoclimatic evidence. *Journal of Human Evolution, 63*(6), 859–866. https://doi.org/10.1016/j.jhevol.2012.08.008.
Clarke, J. G. D. (1969). *World prehistory: A new outline*. Cambridge, UK: Cambridge University Press.
Coon, C. S. (1965). *The living races of man*. New York: Alfred A. Knopf.
Dart, R. A. (1923). Boskop remains from South-East coast. *Nature, 112*, 623–625.
Darwin, C. (1859). *On the origin of species, or the preservation of favoured races in the struggle for life*. London: John Murray, Albemarle Street.
de Maret, P. (2013). Archaeologies of the bantu expansion. In P. Mitchell, & P. J. Lane (Eds.), *The Oxford handbook of African archaeology* (pp. 627–644). Oxford: Oxford University Press.
d'Errico, F., Backwell, L., Villa, P., Degano, I., Lucejko, J. J., Bamford, M. K., … Beaumont, P. B. (2012). Early evidence of San material culture represented by organic artifacts from Border Cave, South Africa. *Proceedings of the National Academy of Sciences of the United States of America, 109*(33), 13214–13219. https://doi.org/10.1073/pnas.1204213109.
Elphick, R. (1977). *Kraal and castle: Khoikhoi and the founding of White South Africa. Vol. 116*. New Haven: Yale University Press.
Fan, S. H., Kelly, D. E., Beltrame, M. H., Hansen, M. E. B., Mallick, S., Ranciaro, A., … Tishkoff, S. A. (2019). African evolutionary history inferred from whole genome sequence data of 44 indigenous African populations. *Genome Biology, 20*(1). https://doi.org/10.1186/s13059-019-1821-1.
FitzSimons, F. W. (1926). Cliff dwellers of Zitzikama: Results of recent excavations. *South African Journal of Science, 23*, 813–817.
Galloway, A. (1937). The characteristics of the skull of the Boskop physical type. *American Journal of Physical Anthropology, 23*, 31–47.
Gordon, R. J. (1997). *Picturing the bushmen: The Denver African expedition of 1925*. Athens, Ohio: Ohio University Press.
Gould, S. J. (1981). *The Mismeasure of man*. New York: W.W. Norton.
Henshilwood, C. S., d'Errico, F., van Niekerk, K. L., Coquinot, Y., Jacobs, Z., Lauritzen, S. E., … Garcia-Moreno, R. (2011). A 100,000-year-old ochre-processing workshop at Blombos Cave, South Africa. *Science, 334*(6053), 219–222. https://doi.org/10.1126/science.1211535.
Lander, F., & Russell, T. (2018). The archaeological evidence for the appearance of pastoralism and farming in Southern Africa. *PLoS One, 13*. https://doi.org/10.1371/journal.pone.0198941.
Legassick, M., & Rassool, C. (2000). *Skeletons in the cupboard: South African museums and the trade in human remains 1907–1917*. Cape Town: South African Museum.
Lewis-Williams, J. D. (2000). Introduction. In J. D. Lewis-Williams (Ed.), *Stories that float from Afar* (pp. 1–41). Cape Town, South Africa: David Philip.
Lombard, M., Wadley, L., Deacon, J., Wurz, S., Parsons, I., Mohapi, M., … Mitchell, P. (2012). South African and Lesotho Stone Age sequence updated. *South African Archaeological Bulletin, 67*(195), 123–144.
Mitchell, P. (2002). *The archaeology of southern Africa*. Cambridge, UK: Cambridge University Press.
Morris, A. G. (1992). *A master catalogue: Holocene human skeletons from South Africa*. Johannesburg, South Africa: Witwatersrand University Press.
Morris, A. G. (2005). Measure by measure: The history of race and typology in South African physical anthropology. In G. Štrkalj, N. Pather, & B. Kramer (Eds.), *Voyages in Science: Essays by South African anatomists in honour of Phillip V. Tobias 80th Birthday* (pp. 121–140). Pretoria: Content Solutions.

Morris, A. G. (2012). Biological anthropology at the southern tip of Africa carrying European baggage in an African context. *Current Anthropology, 53*, S152–S160. https://doi.org/10.1086/662289.

Morris, A. G. (2014). Controversies about the study of human remains in post-apartheid South Africa. In B. Odonnabhain, & M. C. Lozada (Eds.), *Archaeological Human Remains: Global Perspectives* (pp. 189–198).

Morris, A. G., Heinze, A., Chan, E. K. F., Smith, A. B., & Hayes, V. M. (2014). First ancient mitochondrial human genome from a prepastoralist southern African. *Genome Biology and Evolution, 6*(10), 2647–2653. https://doi.org/10.1093/gbe/evu202.

Pargeter, J., MacKay, A., Mitchell, P., Shea, J., & Stewart, B. A. (2014). Primordialism and the 'Pleistocene San' of southern Africa. *Antiquity*, 1072–1079. https://doi.org/10.15184/aqy.2016.100.

Penn, N. G. (1996). "Fated to perish": The destruction of the Cape San. In P. Skotnes (Ed.), *Miscast: Negotiating the presence of the bushmen* (pp. 81–91). Cape Town: UCT Press.

Pfeiffer, S. (2012). Conditions for evolution of small adult body size in southern Africa. *Current Anthropology, 53*(S6), S383–S394.

Rightmire, G. P. (1978). Human skeletal remains from the southern Cape Province and their bearing on the Stone Age prehistory of South Africa. *Quaternary Research, 9*, 219–230.

Rightmire, G. P., & Deacon, H. J. (1991). Comparative studies of Late Pleistocene human remains from Klasies River Mouth, South Africa. *Journal of Human Evolution, 20*, 131–156.

Sadr, K. (2013). A short history of early herding in Southern Africa. In M. Bollig, M. Schnegg, & H.-P. Wotzka (Eds.), *Pastoralism in Africa: Past, present and futures* (pp. 171–197). New York and Oxford: Berghahn Books.

Schlebusch, C. M., Malmström, H., Günther, T., Sjödin, P., Coutinho, A., Edlund, H., … Jakobsson, M. (2017). Southern African ancient genomes estimate modern human divergence to 350,000 to 260,000 years ago. *Science, 358*, 652–655.

Schlebusch, C. M., Prins, F., Lombard, M., Jakobsson, M., & Soodyall, H. (2016). The disappearing san of southeastern Africa and their genetic affinities. *Human Genetics, 135*(12), 1365–1373. https://doi.org/10.1007/s00439-016-1729-8.

Singer, R. (1958). The Boskop 'race' problem. *Man, 232*, 173–178.

Skoglund, P., Thompson, J. C., Prendergast, M. E., Mittnik, A., Sirak, K., Hajdinjak, M., … Reich, D. (2017). Reconstructing prehistoric African population structure. *Cell, 171*(1), 59. https://doi.org/10.1016/j.cell.2017.08.049.

Stynder, D. D., Ackermann, R. R., & Sealy, J. C. (2007a). Craniofacial variation and population continuity during the South African Holocene. *American Journal of Physical Anthropology, 134*, 489–500.

Stynder, D. D., Ackermann, R. R., & Sealy, J. C. (2007b). Early to mid-Holocene South African Later Stone Age human crania exhibit a distinctly Khoesan morphological pattern. *South African Journal of Science, 103*, 349–352.

Tobias, P. V. (1959). Epilogue: Some developments in South African physical anthropology 1938–1958. In A. Galloway (Ed.), *The skeletal remains of Bambandyanalo* (pp. 129–154). Johannesburg: Witwatersrand University Press.

Tobias, P. V. (2002). Saartje Baartman: Her life, her remains, and the negotiations for their repatriation from France to South Africa. *South African Journal of Science, 98*(3–4), 107–110. Retrieved from <Go to ISI>://WOS:000176435700002.

Wolpoff, M., & Caspari, R. (1997). *Race and human evolution*. New York: Simon and Schuster.

Wood, B. (2019). *Human evolution: A very short introduction* (2nd ed.). Oxford, UK: Oxford University Press.

CHAPTER 10

Southern African foragers: Life histories of past communities

Approaching the question of body size

This chapter describes my long quest to learn about being short. The story line follows my own program of research, organized around the questions that struck me as most important. You will see many other names in the citations. They are former students and colleagues whose work complemented my own, and without whom I would still be spinning my wheels. As you scan the citations, you will see frequent reference to the Later Stone Age (see Box 9.2). That is the archeologically defined, cultural context associated with the burials. For decades it has been considered good practice in archaeology not to link a cultural pattern to a particular group of people. In this case, though, the linkage between the archeological and genomic data is so strong that I can refer to the people as ancestral Khoe-San. This is the back-story of one branch of a lineage that lives on, albeit in a different part of Southern Africa.

Most people are probably not aware how often they think about height, a.k.a. stature. Relative body size is one of the first physical features we note about a new acquaintance. Parents anxiously monitor the linear growth of their offspring, marking progress on a wall. Sportscasters comment on athletes' stature, whether they are describing jockeys, basketball players, cricketers, or competitive divers. A person's stature is correlated with all manner of life outcomes. Height is highly correlated with labor market success in industrialized countries (Case & Paxson, 2008). Taller men fare better in marriage markets (Murasko, 2020). This is to be expected, since the very common global practice of hypergamy (in which a woman marries "up") has three components: education, age and stature. Being a tall man is a good thing. In the marriage market, a woman benefits from being medium-sized, although tallness is to her professional advantage. While there is not likely to be a simple explanation for why we like to "look up to" those who are in charge, there are a lot of social phenomena reinforcing our bias. There is a line of scholarship purporting that height is positively associated with

Osteobiographies
https://doi.org/10.1016/B978-0-12-823880-6.00011-2

cognitive ability (Case & Paxson, 2008). Studies in education report measurable damage from teachers' negative assumptions about the abilities of shorter children (Black & de New, 2020). To confirm the pervasiveness of this bias, just do an on-line search for short people jokes, short people songs and memes. This bias—favoring the tall and doubting the abilities of the short—is not new.

Height is a classic polygenic trait (Stulp & Barrett, 2016). It reflects the influence of thousands of variable positions across the genome. Height differences between Northern and Southern Europeans have been shown to be the result of selection for tallness among Northern Europeans (Turchin et al., 2012). In the lingo of academics, there is positive selection for stature through life history processes (Stulp & Barrett, 2016). In much of the world height is destiny, and more is better.

In Southern Africa, since about C.E. 1652, newcomer Europeans disdained the petite Khoe-San aboriginals, and assumed them to be weak and frail, perhaps the result of malnutrition or inbreeding depression. A 1927 magazine caption for a photo of a European with two Bushmen read, "to offset the handicap of small stature, in a country of fierce struggle for existence, where only the strong or crafty survive, the Bushman of the African wilds has developed, through the decades, traits that have made his name a terror to Europeans" (Gordon, 1997:166). Within this contested landscape it was assumed that smallness was a disadvantage. Was it? Was smallness a transient phenomenon reflecting difficulties of the times, or an established characteristic of the population?

Historic sources consistently reported that Khoe-San peoples were small in body size. Early European writers noted that the Khoikhoi pastoralists from whom they got livestock were about 10 cm taller than the San foragers, yet in facial shape, hair and skin features the two communities looked very similar. A popular idea was that the San's smallness might reflect food shortage: a health problem to be corrected (Tobias, 1962; Wells, 1960). A group of early 20th century San living in the Kalahari region (Dart, 1937) had mean statures of 61.5 in./155.8 cm (men) and 57.5 in./146.1 cm (women). Decades later, the Harvard Kalahari study found mean statures of 63.3 in./160.9 cm (men) and 59 in./150 cm (women) (Truswell & Hanson, 1976). To put these numbers into perspective, only 5% of Americans are shorter than 163 cm (men) and 150 cm (women) (National Center for Health Statistics, 2016). Truswell and Hanson reported weights for the community of 105.6 lb./47.9 kg (men) and 88.4 pounds/40.1 kg (women). These values are even further below the contemporary American 5th percentiles of 62 kg (men) and 50 kg (women) (National Center for Health Statistics, 2016). These numbers illustrate that the San were not only short, but also very lean.

Genomic studies have documented a substantial divergence within the Khoe-San lineage, about 35,000 years ago, into northern and southern language groups (Schlebusch et al., 2012). The Kalahari communities are members of the northern language group, and the curated skeletons almost certainly represent the southern group (Skoglund et al., 2017). Would expectations based on parallels with the Kalahari ethnography be met? The upper leg bone—the femur—makes a notable contribution to a person's stature. Among the many tools we have for estimating body dimensions in life, the femur length to stature relationship is one of the best. Using a formula that included some Khoe-San information in its creation (Feldesman & Fountain, 1996), a picture of stature across time and space can be developed from maximum femur length measurements (Fig. 10.1). The estimated statures of the skeletons are consistent with those from their living relatives (Kurki, Ginter, Stock, & Pfeiffer, 2010). There is no evidence that

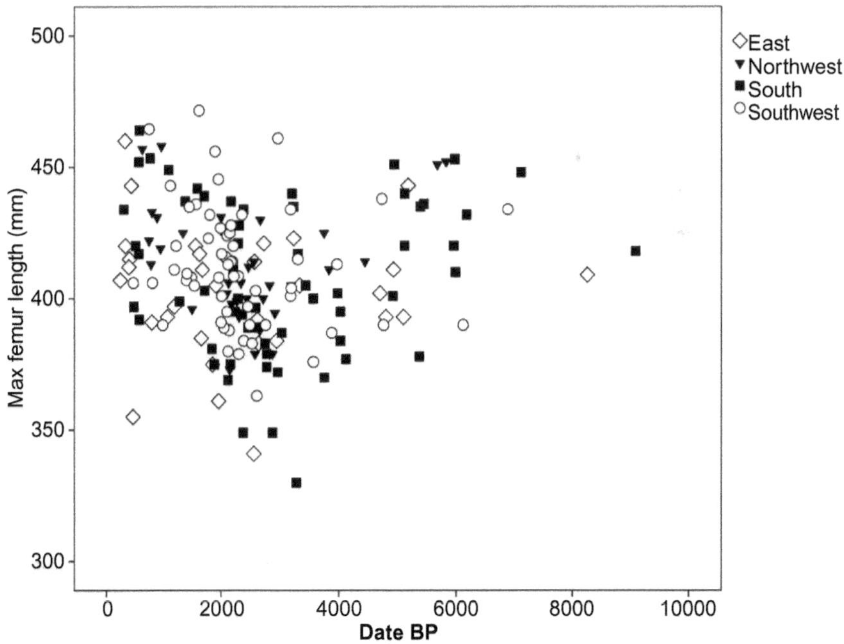

Fig. 10.1 A plot of maximum femur lengths from 185 ancestral Khoe-San skeletons with radiocarbon dates, labeled by geographic sectors (Pfeiffer, 2016). All except the Northeast are within the fynbos region. All regions show a similar pattern.

there was a taller population in the region at any time during the Holocene. There is a notable number of very, very short people during the period of roughly 3000 to 2000 years ago. This aligns with the period from which we have the most skeletons (Fig. 9.3). While the discovery of very small femora only within this window of time may reflect chance alone, it may also be a real anomaly that needs to be considered as the story moves along.

Reconstructions of body mass are less accurate than reconstructions of stature, but they can be helpful. Whenever a person stands or walks, much of their body mass is transmitted through their femur. The diameter of the head of the femur (the ball of the hip socket) corresponds to body mass (McHenry, 1992; Ruff, Scott, & Liu, 1991). The width of the pelvis is an even better basis for mass estimation, but survival of all three bones of the pelvic girdle is much less common than survival of the femoral head. Using all methods, the estimates of body mass of the people for whom we have skeletons line up very closely with expectations: about 40–50 kg (Kurki et al., 2010). Extended joint capsules of the bones of the ankle provide more evidence for leanness. They indicate that a deep squat was a common habitual posture (Dewar & Pfeiffer, 2004) (Fig. 10.2). This is a posture that is impossible if one's legs are not lean. So, the skeletons show that smallness in both stature and mass was a long-standing ancestral Khoe-San trait. But what does that mean? Does it mean that they were chronically deprived, undernourished, diseased? To answer these questions, we needed other lines of evidence. While each topic would have been interesting on its own, the population's uniquely small body size influenced the specific questions we asked (Fig. 10.3).

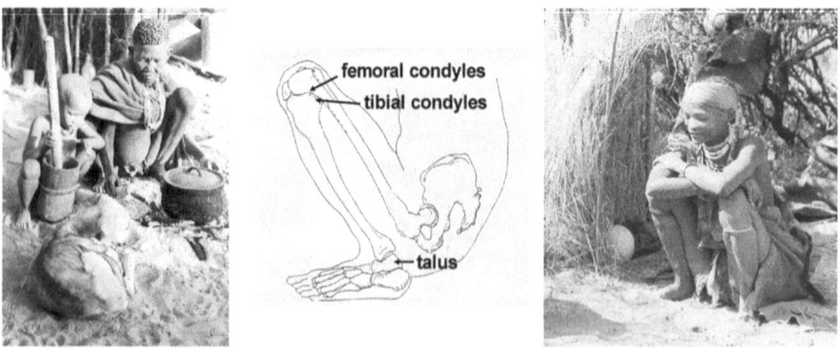

Fig. 10.2 A diagram in the center shows points at which the joint capsules can become extended when deep dorsiflexion is a frequent posture; on both sides, San people assuming habitual deep flexion postures. They are members of the Dobe!Kung (Ju/'hoan) community (Lee, 2003), photographed by R. B. Lee in 1967. *(With permission of R.B. Lee.)*

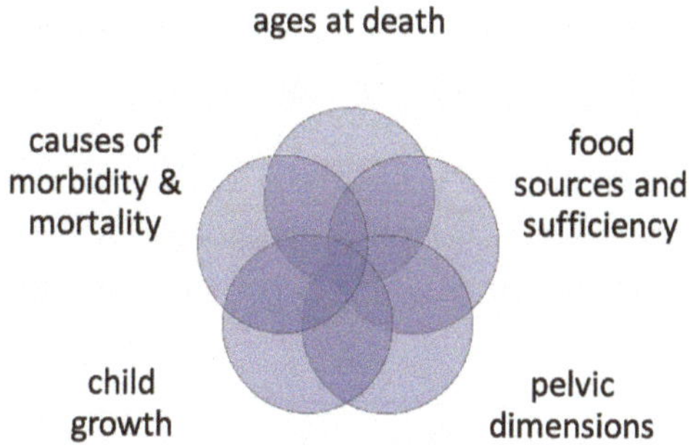

Fig. 10.3 The perspectives that contribute to understanding persistently small adult body size.

What did people eat?

The fynbos biotic province is home to a remarkable array of plants providing food, as well as medicinal and plants for crafting (van Wyk & Gericke, 2000; van Wyk, van Outshoorn, & Gericke, 1997). Particularly important as food are the plants that grow underground storage organs in the form of roots and corms, especially from the plant family Iridaceae (Marean et al., 2015). Field experiments in relatively undisturbed areas indicate that lots of food energy (calories) could be gathered in relatively little time, although there would have been issues of seasonal availability (De Vynck, Cowling, Potts, & Marean, 2016). There are also leafy greens and fruiting plants. Terrestrial animals in the past included a broad array, but it seems that Khoe-San foragers focused on small game, ostrich eggs, small mammals, snakes and tortoises (Deacon & Deacon, 1999; Steele & Klein, 2013). Commonly found in occupational contexts since Middle Stone Age times, tortoises provided protein, fat and calories, available after relatively brief procurement and preparation times (Thompson & Henshilwood, 2014).

Foraging along the shorelines could provide diverse sources of protein, ranging from shellfish at the lower end of the food chain to occasional seal and whale wash-ups at the upper end. Shellfish appear to have been especially important. Shellfish provide relatively large amounts of protein, fat and energy (Venugopal & Gopakumar, 2017). They yield fat in greater proportion than is available from the muscle meat of terrestrial nondomesticated animals (Buchanan, 1988; Kyriacou, 2017). However, the availability of the intertidal zone from which shellfish can be harvested is limited by lunar cycles and by seasonally rough seas (De Vynck, Anderson, et al., 2016). There is no evidence that Khoe-San people used boats or practiced deep-water diving. The South Atlantic Ocean is very cold. Both it and the

Indian Ocean have very strong currents and sharks. This was a coastal, not a maritime, adaptation (Marean et al., 2014).

Some information about peoples' diets can be obtained from stable isotope studies of their bones and teeth. Those studies have provided evidence about the broad categories of food that comprised the diet. The pattern can then be interpreted in the context of archeological and ecological evidence for available foodstuffs (Pfeiffer & Sealy, 2006; Sealy, 2010; Sealy & Pfeiffer, 2000). Studies like these support the generalization that shellfish were an important source of protein. One of the most interesting dietary isotope studies shows that in a section of coast where there is a mainland seal rookery (which is very unusual), people near this source of desirable protein somehow denied their neighbors access (Sealy, 2006). We have very little evidence of the foragers' home bases, and scant evidence of food processing or storage. It appears that there were sources of high-quality food. People needed to have fine grained knowledge of the landscape and they needed to be prepared to move frequently among resources. This latter point is consistent with the evidence from biomechanical measures of femur strength. The Khoe-San ancestors have exceptionally strong femora, compared to past peoples from other parts of the world (Cameron & Stock, 2018; Stock & Pfeiffer, 2001). This is not to say that they moved across different, distinct ecozones. The stable isotope values line up with local resources. For example, people who were buried at a substantial distance from the coast have stable isotope ratios consistent with terrestrial protein, not seafood (Sealy, Patrick, Morris, & Alder, 1992). The developing picture may be one of hunter-gatherer societies quite commonly maintaining territories, at least to some extent.

What about the children?

Was there enough food? The question of whether there was normally enough food to sustain child growth, health and reproduction must be approached from perspectives other than stable isotopes. One approach is to look for the sort of abnormal development of bones and teeth that reflect what bioarchaeologists call "chronic stress indicators." Because a growing child is particularly sensitive to stressors like food shortages, infectious and parasites, their skull bones and tooth crowns sometimes reflect responses to these stresses. Porous bone of the orbits (called *cribra orbitalia*) and on the skull vault (called porotic hyperostosis), as well as defects of tooth enamel (hypoplasia, see Box 2.1) are often assessed. These features can show up as slight to quite severe. In addition, radiographs of juvenile long bones will show lines perpendicular to the long axis if growth has been arrested and then started up again. Surveys of ancestral Khoe-San children who failed to survive have noted that all these developmental defects are infrequent, relative to comparators from elsewhere. A survey of 43 juveniles found 13 with slight to moderate *cribra orbitalia* and none with porotic hyperostosis. There were very few indications of growth arrest lines (Pfeiffer, 2007). It is a bit difficult to study enamel hypoplasias because the chewing surfaces of peoples' teeth became worn from the grit in their diets. A study of the permanent tooth crowns of a small number of children shows evidence of minor interruptions to growth during early childhood (Harrington & Pfeiffer, 2016; Pfeiffer & Harrington, 2018). In sum, we do not see skeletal indications of stressful lives. On the other hand, some children failed to survive. What could have interrupted their lives?

Tracking child growth

It is a sad fact of modern life that forensic investigators need good data on the dimensions of newborn babies' long bones. When very small infants are found under suspicious circumstances, forensic scientists need to distinguish the developmental age of the remains. Bioarchaeologists can use these data to estimate the gestational status of infants who failed to survive. As noted in earlier chapters, the best age estimates come from dental evidence, even at birth (Box 2.1). When femur length is plotted against dental age in ancestral Khoe-San neonates, their sizes are similar to those of European newborns. This lines up with information from prolonged Kalahari fieldwork, where among ten births, only one was below the World Health Organization low birth weight criterion of 2500 g (Howell, 2010). Two questions arise from this: How do petite women give birth to relatively large babies? If the babies are European size at birth, when does the trend toward smallness begin?

Taking the second question first, we can use modern data sets that characterize normal linear growth of healthy children to frame an answer. Normal child growth has a distinct tempo, with spurts of growth in the first two years of life, a slower tempo in mid-childhood, then another spurt of growth at puberty. Girls reach puberty earlier than boys, and boys have a stronger pubescent growth spurt than girls. These sex differences muddy the picture somewhat when data are combined for boys and girls. Nevertheless, the uptick in late childhood is clear. Using ages estimated from dental development, we can plot the lengths of children's long bones to create a cross-sectional picture of ancestral Khoe-San child growth. When we did this, we found that the shape of the curve looks normal, but the pitch is subdued. If we plot Khoe-San absolute values against values from North American children, all Khoe-San children look stunted. But if we plot those values as the proportion of adult size we would expect by a particular age, they are on-target, with relatively little variability (Fig. 10.4). From this line of investigation, we can conclude two important things. First, the Khoe-San children who failed to survive were following a normal tempo of growth toward a petite set-point (Pfeiffer & Harrington, 2011). There is little evidence of individual children falling behind to varying extents, as we would see if growth faltering was a common phenomenon. Second, this normal tempo suggests that causes of death were acute rather than chronic. There is very little evidence that any child fell ill and held on to life for a considerable time before dying. We often see a pattern of substantial growth lag among children who died young in communities affected by endemic infectious diseases. Ancestral Khoe-San children appear to have died of things that were immediately fatal, like animal attacks, drowning, burns, snake bites, and the like. While we cannot know causes of death, we can at least tentatively rule out infectious diseases as common.

Pelvic dimensions and childbirth

How do petite women give birth to relatively large babies? Perhaps being a small adult had some advantages, but surely it could have at least one big disadvantage. Women need to successfully give birth to babies who are mature enough to survive. The relatively tight fit between the obstetric canal and the neonate is a topic of great interest in human evolution, often referred to as the "obstetric dilemma" (Wall-Scheffler, Kurki, & Auerbach, 2020). The pelvis must be narrow enough to be efficient in locomotion, yet not so narrow as to jeopardize reproduction. If the infant cannot fit through the birth canal at the moment of birth (parturition), the result may be death for mother, infant, or both. This is an anatomical feature on which natural selection acts quite directly.

Thanks to excellent preservation of the bones that make up the pelvis, we can study rather large numbers of ancestral Khoe-San adult skeletons, to get a strong sense of population characteristics. The external dimensions of their pelves match what would be expected for their body size, but the lower planes of the pelvic (obstetric) canal are elongated front to back. The slightly different shape of the obstetric canal benefits clearance at parturition (Kurki, 2007, 2013). This shape is unique. If young women died during childbirth, or if their infants died, this would select against the anatomical features of those women's skeletons. Given enough time, this is how natural

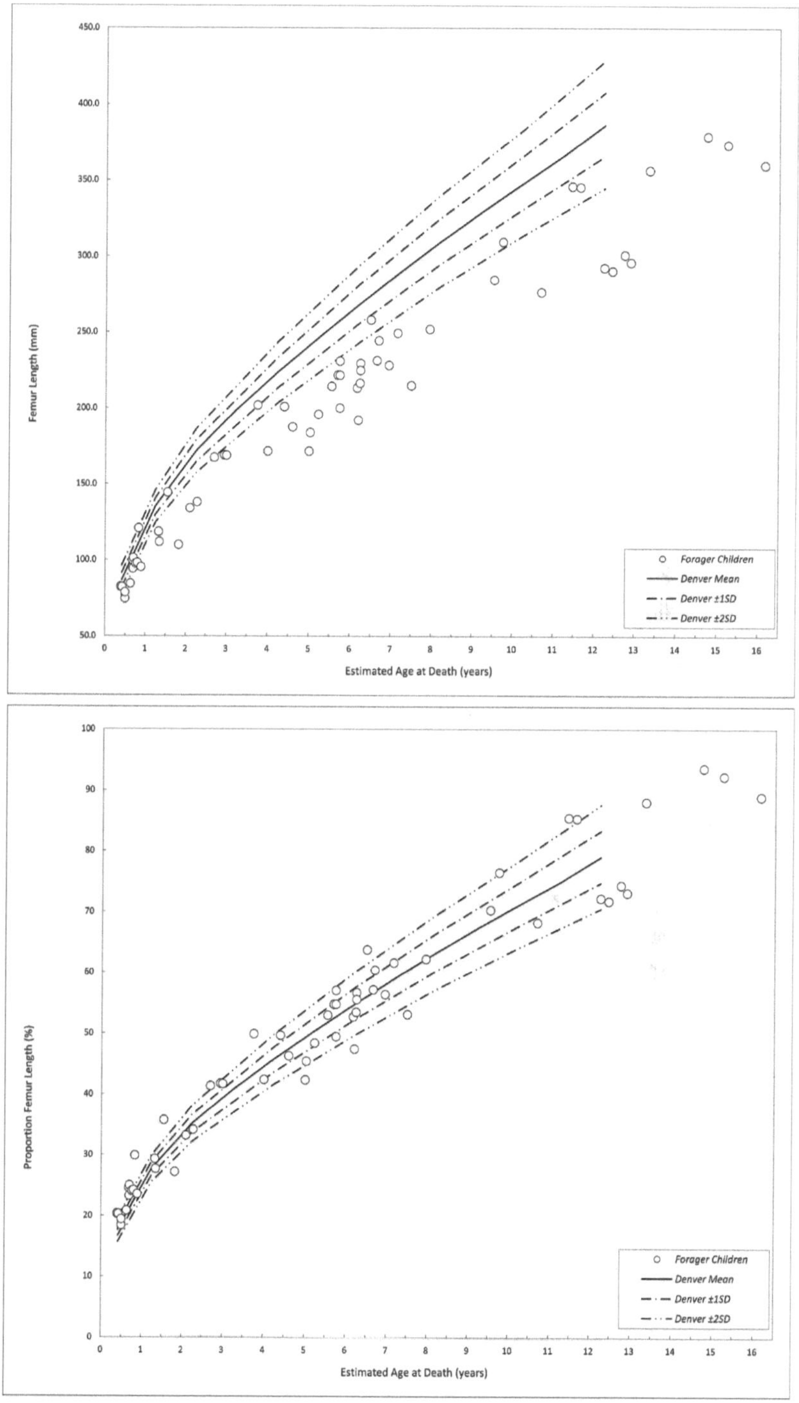

Fig. 10.4 Femur lengths of ancestral Khoe-San children plotted against North American norms (Maresh, 1970), in two ways: Above, absolute values for femur length, by age of child; below, values of femur length relative to the expected acquisition of adult size. *(Credit: L. Harrington.)*

selection could cause a population to have a particular, adaptive pelvic shape. There is also some direct evidence of women's reproductive health challenges. The skeleton of one older woman shows extensive osteoarthritis at all the joints of the pelvic girdle—both at the front (pubic symphysis) and at the back (sacro-iliac joints). This is very unusual. It may have been caused by trauma during parturition, stretching the normally stiff attachment sites beyond a point from which they could recover. Muscle attachment sites on the pelvis and legs indicate that the way she walked was affected by her loose—and probably painful—joints (Pfeiffer, 2011). She lived for many years after the birth event, and we can hope that the infant did, too.

Illness, accidents and survival

As noted earlier, the children who failed to survive seemed to be growing well, and the signs of poor health that show up on bones and teeth are uncommon and modest in their extent. The same is true of the adult skeletons. There is no clear evidence of recovery from bouts of fever or diarrhea, as we might expect if diseases like malaria and cholera were affecting the communities. There is also no evidence, at least as yet, of the bone shape changes associated with *Mycobacterium tuberculosis*, nor with infectious agents like *Staphylococcus*. Blood-born staph infections can affect bone, leading to pathological remodeling (Waldron, 2009). Among the several hundred skeletons, there are instances of inherited disease that shortened life (Pfeiffer & Crowder, 2004), various well healed long bone fractures and cranial wounds from which people recovered. In most instances, there are no clues about what caused a person to die. Aspects of this topic are discussed further in Chapter 11.

Life expectancy

Whole genome studies of living Khoe-San people show not only that the lineage was isolated for a long time, but also that the effective population size was quite large, with no evidence of periodic declines in numbers (Kim et al., 2014). Children had a good chance of living long enough to reproduce, nurture their offspring and perhaps help to nurture their grandchildren. This aligns with contemporary demographic evidence. Extended observations of northern Khoe-San foraging communities document that older adults made up a notable proportion of local groups. Of the several hundred Ju/'hoan-!Kung studied in the late 1960's, about 8% were more than 60 years old. Most of the old folks were in robust good health (Howell, 2000). There must have been postreproductive (elderly) members in past human societies (Crews & Stewart, 2010; Kim, McQueen, & Hawkes, 2019). At least some ancestral Khoe-San skeletons should show evidence of considerable longevity.

This was not immediately apparent. The methods used for estimating age at death for adults older than about 25 years tend to rely on wear-and-tear: the breakdown of tissue around joint surfaces and the stiffening of connective tissue. Although we could readily identify those who had died as young adults by using age markers of late-fusing bone growth plates, almost all of the older adults all appeared to have died in what bioarchaeologists call the "mid adult" stage, 35–50 years. Even those who could be placed in the "old adult" category of 50+ years looked like they'd barely made it to that ripe old age. Where were the grandparents? The scenario of so many people dying young is inconsistent with expectations. Could it instead mean that the skeletons of small, lean adults do not show the characteristic age-related bone changes? Yes, it could. In a large study of skeletons in modern anatomical collections whose stature and weight in life are known, Catherine Merritt has demonstrated that light weight, small-bodied adults are consistently under-aged when the most commonly used age estimation methods are used (Merritt, 2015, 2017). We applied a different age estimation method—one based on histological (microscopic) assessment of rib bone cross-sections—to a study of 46 adult ancestral Khoe-San skeletons. The histological method assesses the normal turnover of bone tissue that we all experience throughout life. It is not influenced by wear-and-tear. Histological results placed a substantially larger number of the skeletons into the 50+ category (Pfeiffer, Cameron, Sealy, & Beresheim, 2019). It is becoming clear that with skeletons of the ancestral Khoe-San, we should place more emphasis on features like spinal compression, low bone mass, tooth wear to the point of tooth loss, and less emphasis on the expected changes to the pelvic joint surfaces and ends of ribs. While there is room for more research on this topic, I like to think that this project pushed back on the "nasty, brutish and short" mindset. Many of the ancestral Khoe-San skeletons were elders within their groups. From a demographic perspective, these were successful communities.

Summary

The skeletons of several hundred ancestral Khoe-San children and adults, people whose deaths spanned several millennia, provide glimpses of a successful foraging population. While we must not forget that the absence of evidence is not evidence of absence, at this time there is no evidence of food shortages or infectious diseases. Prior to the arrival of Europeans, there is no evidence of tall people in the fynbos world. There is evidence that unique pelvic shape accommodated safe births of relatively big babies. A larger obstetric canal could have been achieved through larger overall body size but that is not what happened. Being small must have been beneficial. Exactly how that worked may remain a mystery, but ethnographically documented hunting strategies based on tracking, persistence and stealth (Lee, 1979; Liebenberg, 2006; Silberbauer, 1981) may offer one clue. There were accidents and difficult life events, but genomic evidence suggests that the foraging lifeway worked pretty well for thousands of years.

However, there are indicators that all was not idyllic. For one thing, there is that time period during which some adults were even shorter than usual (Fig. 10.1). For another thing, there are some skeletons that show signs of having been killed by other humans. Those instances generally correspond with the period of shorter stature, but not neatly. The question of when and how interpersonal violence played a role in the lives of Holocene foragers of southern-most Africa is the topic of the next chapter.

References

Buchanan, W. F. (1988). *Shellfish in prehistoric diet, Elands Bay, S.W. Cape Coast, South Africa.* Oxford: Cambridge University Press.

Black, N., & de New, S. C. (2020). Short, heavy and underrated? Teacher assessment biases by children's body size. *Oxford Bulletin of Economics and Statistics.* https://doi.org/10.1111/obes.12370.

Cameron, M. E., & Stock, J. T. (2018). Ecological variation in Later Stone Age southern African biomechanical properties. *Journal of Archaeological Science: Reports, 17,* 125–136.

Case, A., & Paxson, C. (2008). Stature and status: Height, ability, and labor market outcomes. *Journal of Political Economy, 116*(3), 499–532. https://doi.org/10.1086/589524.

Crews, D. E., & Stewart, J. A. (2010). Human longevity and senescence. In M. P. Muehlenbein (Ed.), *Human evolutionary biology* (pp. 528–550). Cambridge: Cambridge University Press.

Dart, R. A. (1937). The physical characters of the /?Auni-/=khomani bushmen: Appendices A-W. *Bantu Studies, 11,* 175–246.

De Vynck, J. C., Anderson, R., Atwater, C., Cowling, R. M., Fisher, E. C., Marean, C. W., … Hill, K. (2016). Return rates from intertidal foraging from Blombos Cave to Pinnacle Point: Understanding early human economies. *Journal of Human Evolution, 92,* 101–115.

De Vynck, J., Cowling, R. M., Potts, A. J., & Marean, C. W. (2016). Seasonal availability of edible underground and aboveground carbohydrate resources to human foragers on the Cape south coast, South Africa. *Peer J, 4.* https://doi.org/10.7717/peerj.1679. e1679.

Deacon, H. J., & Deacon, J. (1999). *Human beginnings in South Africa.* Cape Town: David Philip Publishers.

Dewar, G., & Pfeiffer, S. (2004). Postural behavior of Later Stone Age people in South Africa. *South African Archaeological Bulletin, 59*(180), 52–58.

Feldesman, M. R., & Fountain, R. L. (1996). "Race" specificity and the femur/stature ratio. *American Journal of Physical Anthropology, 100,* 207–224.

Gordon, R. J. (1997). *Picturing the Bushmen: The Denver African expedition of 1925*. Athens, Ohio: Ohio University Press.

Harrington, L., & Pfeiffer, S. (2016). Developmental variation in perikymata expression in co-interred child foragers. *American Journal of Physical Anthropology, 159*(S62), 168.

Howell, N. (2000). *Demography of the Dobe !Kung* (2nd ed.). New York: Aldine de Gruyter.

Howell, N. (2010). *Life histories of the Dobe !Kung: Food, fatness and well-being over the lifespan*. Berkeley: University of California Press.

Kim, P. S., McQueen, J. S., & Hawkes, K. (2019). Why does women's fertility end in mid-life? Grandmothering and age at last birth. *Journal of Theoretical Biology, 461*, 84–91.

Kim, H. L., Ratan, A., Perry, G. H., Montenegro, A., Miller, W., & Schuster, S. C. (2014). Khoisan hunter-gatherers have been the largest population throughout most of modern-human demographic history. *Nature Communications, 5*(5692). https://doi.org/10.1038/ncomms6692.

Kurki, H. K. (2007). Protection of obstetric dimensions in a small-bodied human sample. *American Journal of Physical Anthropology, 133*, 1152–1165.

Kurki, H. K. (2013). Bony pelvic canal size and shape in relation to body proportionality in humans. *American Journal of Physical Anthropology, 151*(1), 88–101. https://doi.org/10.1002/ajpa.22243.

Kurki, H. K., Ginter, J., Stock, J., & Pfeiffer, S. (2010). Body size estimation of small-bodied humans: Applicability of current methods. *American Journal of Physical Anthropology, 141*(2), 169–180.

Kyriacou, K. (2017). Intertidal shellfish as a source of protein and energy for the Middle Stone Age inhabitants of the southwestern Cape and northern KwaZulu-Natal, South Africa. *Quaternary International, 438*(B), 30–39.

Lee, R. B. (1979). *The !Kung san: Men, women and work in a foraging society*. Cambridge: Cambridge University Press.

Lee, R. B. (2003). *The Dobe Ju/'hoansi*. New York: Harcourt Brace.

Liebenberg, L. (2006). Persistence hunting by modern hunter-gatherers. *Current Anthropology, 47*, 1017–1026.

Marean, C. W., Anderson, R. J., Bar-Matthews, M., Braun, K., Cawthra, H. C., Cowling, R. M., … Zahn, R. (2015). A new research strategy for integrating studies of paleoclimate, paleoenvironment, and paleoanthropology. *Evolutionary Anthropology: Issues, News and Reviews, 24*(2), 62–72.

Marean, C. W., Cawthra, H. C., Cowling, R. M., Esler, K. J., Fisher, E., Milewski, A., … De Vynck, J. (2014). Stone age people in a changing south African greater cape floristic region. In N. Allsopp, J. F. Colville, & G. A. Verboom (Eds.), *Fynbos: Ecology, evolution, and conservation of a megadiverse region* (pp. 164–199). Oxford, UK: Oxford University Press.

Maresh, M. M. (1970). Measurements from roentgenograms. In R. W. McCammon (Ed.), *Human growth and development* (pp. 157–200). Springfield: Charles C. Thomas.

McHenry, H. M. (1992). Body size and proportions in early hominids. *American Journal of Physical Anthropology, 87*, 407–431.

Merritt, C. E. (2015). The influence of body size on adult skeletal age estimation methods. *American Journal of Physical Anthropology, 156*, 35–57.

Merritt, C. E. (2017). Inaccuracy and bias in adult skeletal age estimation: Assessing the reliability of eight methods on individuals of varying body sizes. *Forensic Science International, 275*. 315.e311-315.e311.

Murasko, J. E. (2020). Height, marriage, and partner characteristics for women in low- and middle-income countries. *Economics and Human Biology, 38*. https://doi.org/10.1016/j.ehb.2020.100876.

National Center for Health Statistics. (2016). *Anthropometric reference data for children and adults: United States, 2011–2014*. Hyattsville, MD: U.S. Government.

Pfeiffer, S. (2007). The health of foragers: People of the Later Stone Age, Southern Africa. In M. N. Cohen, & G. Crane-Kramer (Eds.), *Ancient health: Skeletal indicators of agricultural and economic intensification* (pp. 223–236). Gainesville: University Press of Florida.

Pfeiffer, S. (2011). Pelvic stress injuries in a small-bodied forager. *International Journal of Osteoarchaeology, 21*(6), 694–703.
Pfeiffer, S., Cameron, M. E., Sealy, J., & Beresheim, A. C. (2019). Diet and adult age-a-death among mobile foragers: A synthesis of bioarchaeological methods. *American Journal of Physical Anthropology, 170*, 131–147.
Pfeiffer, S., & Crowder, C. (2004). An ill child among mid-Holocene foragers of Southern Africa. *American Journal of Physical Anthropology, 123*(1), 23–29.
Pfeiffer, S., & Harrington, L. (2011). Bioarchaeological evidence for the basis of small adult stature in Southern Africa: Growth, mortality and small stature. *Current Anthropology, 52*(3), 449–461.
Pfeiffer, S., & Harrington, L. (2018). Regional continuity and local distinctions among Holocene foragers of coastal South Africa: Defining identity at different scales. In D. Temple, & C. M. Stojanowski (Eds.), *Hunter-gatherer resilience in bioarchaeological perspective* (pp. 26–46). Cambridge, UK: Cambridge University Press.
Pfeiffer, S., & Sealy, J. (2006). Body size among Holocene foragers of the Cape ecozone, Southern Africa. *American Journal of Physical Anthropology, 129*, 1–11.
Pfeiffer, S. (2016). An exploration of interpersonal violence among Holocene foragers of Southern Africa. *International Journal of Palaeopathology, 13*, 27–38.
Ruff, C. B., Scott, W. W., & Liu, A.Y. (1991). Articular and diaphyseal remodeling of the proximal femur with changes in body mass in adults. *American Journal of Physical Anthropology, 86*, 397–413.
Schlebusch, C. M., Skoglund, P., Sjodin, P., Gattepaille, L. M., Hernandez, D., Jay, F., … Jakobsson, M. (2012). Genomic variation in seven Khoe-san groups reveals adaptation and complex African history. *Science, 338*(6105), 374–379. https://doi.org/10.1126/science.1227721.
Sealy, J. C. (2006). Diet, mobility and settlement pattern among Holocene hunter-gatherers in southernmost Africa. *Current Anthropology, 47*(4), 569–595.
Sealy, J. C. (2010). Isotopic evidence for the antiquity of cattle-based pastoralism in southernmost Africa. *Journal of African Archaeology, 8*(1), 65–81.
Sealy, J. C., Patrick, M. K., Morris, A. G., & Alder, D. (1992). Diet and dental caries among Later Stone Age inhabitants of the Cape Province, South Africa. *American Journal of Physical Anthropology, 88*, 123–134.
Sealy, J. C., & Pfeiffer, S. (2000). Diet, body size and landscape use among Holocene peoples in the Southern Cape, South Africa. *Current Anthropology, 41*(4), 642–655.
Silberbauer, G. (1981). *Hunter and habitat in the Central Kalahari Desert*. Cambridge: Cambridge University Press.
Skoglund, P., Thompson, J. C., Prendergast, M. E., Mittnik, A., Sirak, K., Hajdinjak, M., … Reich, D. (2017). Reconstructing prehistoric African population structure. *Cell, 171*(1), 59. https://doi.org/10.1016/j.cell.2017.08.049.
Steele, T. E., & Klein, R. G. (2013). The MIddle and Later Stone Age faunal remains from Diepkloof Rock Shelter, Western Cape, South Africa. *Journal of Archaeological Science, 40*(9), 3453–3462.
Stock, J., & Pfeiffer, S. (2001). Linking structural variability in long bone diaphyses to habitual behaviors: Foragers from the southern African Later Stone Age and the Andaman Islands. *American Journal of Physical Anthropology, 115*(4), 337–348.
Stulp, G., & Barrett, L. (2016). Evolutionary perspectives on human height variation. *Biological Reviews, 91*(1), 206–234.
Thompson, J. C., & Henshilwood, C. S. (2014). Nutritional values of tortoises relative to ungulates from the Middle Stone Age levels at Blombos Cave, South Africa: Implications for foraging and social behavior. *Journal of Human Evolution, 67*, 33–47.
Tobias, P.V. (1962). On the increasing stature of the Bushmen. *Anthropos, 57*, 801–810.

Truswell, A. S., & Hanson, J. D. L. (1976). Medical research among the !Kung. In R. B. Lee, & I. DeVore (Eds.), *Kalahari Hunter-Gatherers* (pp. 166–194). Cambridge, Massachusetts: Harvard University Press.

Turchin, M. C., Chiang, C. W. K., Palmer, C. D., Sankararaman, S., Reich, D., Hirschhorn, J. N., & Genetic Invest, A. (2012). Evidence of widespread selection on standing variation in Europe at height-associated SNPs. *Nature Genetics*, *44*(9), 1015. https://doi.org/10.1038/ng.2368.

van Wyk, B., & Gericke, N. (2000). *People's plants: A guide to useful plants of Southern Africa*. Pretoria: Briza Publications.

van Wyk, B., van Outshoorn, B., & Gericke, N. (1997). *Medicinal plants of South Africa*. Pretoria, S.A.: Briza Publications.

Venugopal, V., & Gopakumar, K. (2017). Shellfish: Nutritive value, health benefits and consumer safety. *Comprehensive Reviews in Food Science and Food Safety*, *16*, 1219–1242.

Waldron, T. (2009). *Palaeopathology*. Cambridge: Cambridge University Press.

Wall-Scheffler, C. M., Kurki, H. K., & Auerbach, B. M. (2020). *The evolutionary biology of the human pelvis: An integrative approach*. Cambridge, UK: Cambridge University Press.

Wells, L. H. (1960). Bushman and Hottentot statures: A review of the evidence. *South African Journal of Science*, *56*, 277–281.

CHAPTER 11

Interpersonal violence among Southern African foragers

Prologue

I live in North America. I have an abiding interest in how hunter-gatherer-foragers managed and thrived, independent of domesticated plants and animals. If we are to understand how this lifestyle was organized—what worked and what didn't—then we need to study locales from around the world. Learning about one's own back yard is necessary, but not sufficient. So, even before there was a common term for it, I was aware of "parachute science" (lately also termed "colonial science"). This is the practice of researchers from high-income countries conducting research that relies on information from low-income countries without affording any benefit to the local scientists or community. In 1992, as I set off to see whether there were curated collections of hunter-gatherer skeletons in South Africa, I began a long and productive collaboration with supportive strangers who became friends. While South Africa thirty years ago was not exactly a low-income country, it was a country that was just emerging from the abuses of apartheid and the isolation of international sanctions. I have tried to be a good and supportive collaborator during the subsequent years, though I can't claim exemplar status. The subject of this chapter provides an excellent point at which to acknowledge the local origins of the ideas herein.

My initial knowledge of Khoe-San peoples was strongly influenced by published ethnographic work. One of the earliest and most popular of these was Elizabeth Marshall's *The Harmless People* (Marshall, 1958). As titled, this account emphasized the peaceful, egalitarian, collaborative nature of Bushman society. As I began studying ancestral Khoe-San skeletons, it was with a focus on life expectancy, challenges to health, and habitual behaviors. I admit that I absorbed the "harmless" idiom into my thinking.

Skeletons from archeological contexts are curated at institutions across South Africa, from Pietermaritzburg in Kwa-Zulu Natal, to Port Elizabeth and Grahamstown in the Eastern Cape, Florisbad and Bloemfontein in the Free State, Kimberley in the Northern Cape Province, and, of course, the

Osteobiographies
https://doi.org/10.1016/B978-0-12-823880-6.00012-4

major cities of Johannesburg and Cape Town. The dispersal of the collections reflects, in part, the pervasive presence of hunter-gatherers across the landscape of southern Africa. I did a lot of traveling, assessing, documenting and measuring catalogued units, aided by Alan Morris's Catalogue (Morris, 1992) and the complementary research interests of Judith Sealy whose priorities were radiocarbon dating and dietary stable isotope studies. While many skeletons had been found as single burials without artifacts or other cultural markers, bone preservation was often excellent, thanks to the dry soil or sand in which they were buried. Once each was dated, the accumulated information from hundreds of individuals could describe a population.

It was quickly apparent that the most numerous collections were at Cape Town institutions: the Faculty of Health Sciences at the University of Cape Town (UCT) and Iziko Museums. The Southwestern Cape region had been a focal region for ancestral Khoe-San peoples, and Cape Town had been an historically important hub for scholarship about them. It was there, in the 1870's, where Wilhelm Bleek, his sister-in-law Lucy Lloyd and his daughter Dorothea Bleek documented/Xam San folklore and language. Those records remain important resources today (Skotnes, 2007).

I was not the first person to study the skeletal collections, some of which date to 19th century excavations. Learning from the curators and other scholars who knew the collections well was crucial to my progress. I have great appreciation for their intellectual generosity. It was my local colleagues who pointed out to me that certain skeletons were distinctive. The ancestral Khoe-San practice had been to bury the deceased alone, flexed on his or her side, accompanied by no material items. When excavators encountered more than one body, or if the body position was askew, these anomalies attracted attention. In those rare cases of two or more skeletons found buried together, excavators had noted possible signs of perimortem trauma (Box 11.1). One of these was a burial found at a place called Quoin Point. It had been published as "Prehistoric Homicide" by UCT colleagues Alan Morris and John Parkington (Morris & Parkington, 1982).

Having expanded my ethnographic reading, I knew that a study of Kalahari Khoe-San groups had documented a comparatively high frequency of men killing one another, usually for reasons of sexual jealousy (Lee, 1979). This research, with its focus on men, was not of much help for explaining the anomalous burials of the Southwestern Cape. The instance described by Morris and Parkington was of a woman who died with projectile points embedded in the bone of her spine. She had been buried with an infant (or perhaps she was in the final stage of pregnancy). Other

BOX 11.1 The response of bone tissue to trauma

This brief summary of how bone responds to externally imposed damage emphasizes aspects that are relevant to questions of interpersonal violence in small-scale societies. Related topics, like bone's response to gun shots, land mines and other modern destructive inventions fall within the purview of forensic anthropology and are not included here.

Bone tissue biology: Bone tissue, in life, is about half mineral and half collagen and water. Its internal organization gives it some flexibility and resistance to external forces. When that resistance fails, fresh bone breaks in ways that reflect its internal structure. This means that a bioarchaeologist can usually tell whether a break occurred while the bone was fresh (either just before or after death: what we call "perimortem"), or was damaged later, after the bone had dried out—while it was buried or when it was excavated (Loe, 2009; Wedel & Galloway, 2014). The breakage pattern of fresh skull bone is particularly distinctive. Depending on its force, blunt force trauma to the vault will cause depression at the point of impact, with in-bending around the edges. If powerful enough to generate radiating fractures, those lines will follow directions of least resistance (for example, they rarely cross sutural lines) and there can be interruptions to the linear breaks (zones of adherence) as the collagen in the bone resists the force (Galloway & Wedel, 2014).

Healing: In life, a freshly broken bone is immediately the focus of repair activities that include increased blood flow and reconstruction of collagen and mineral. A temporary "weld" known as a callus is formed. The callus is gradually remodeled into structurally sound bone. Compared to the healing of soft tissue wounds, this is a slow process. The specifics of healing are dictated in part by which bone it is. The cranial vault—the bone protecting the brain—is diploic: two thin layers of dense bone held together by a thin, spongy layer. When diploic cranial bone is perforated, the healing process will not rebuild diploic bone. A hole with smooth edges will remain (Rao, 2003).

Patterns and causes: Any part of the skeleton can be damaged through accidents and misadventures. Certain kinds of healed breaks to the postcranial skeleton are often seen in skeletons from the past. They are normally well healed, without apparent infection or much displacement. As in modern times, common locations include the collarbone (clavicle), the ribs, and either of the bones of the forearm (radius, ulna). Healed breaks to major weight-bearing bones like the femur and tibia, or to the bones of the pelvis/hip are much less commonly seen in archeologically-derived skeletons. This is probably because ancient skeletons are rarely osteoporotic and because a traumatic event of that magnitude could be fatal, through shock and loss of blood.

While any postcranial damage can be associated with interpersonal violence, most scholarship focuses on the cranium. A broken nasal bone may result from

Continued

BOX 11.1, cont'd

an accident, but more extensive breakage to facial bones is often interpreted as a signal interpersonal conflict. There is also a hypothesis that traumatic damage to cranial bone above where the brim of a hat would sit is not likely to result from an accident. Forensic exploration of the hat-brim line rule shows that damage below the line is associated with falls, whereas homicidal blows lead to damage above the line (Kremer, Racette, Dionne, & Sauvageau, 2008). This maxim is not necessarily applicable to contexts in which people spent their time in rock shelters like those of the Cape Fold Belt, where rock fragments can spall off the overhang (Fig. B11.1).

Fig. B11.1 (A) The top of a man's skull (viewed from the right side), dated to about 3000 years ago, shows partial healing from a single blow with radiating cracks. The roughened bone surface at the point of impact shows that healing was incomplete at the time of his death (Pfeiffer, 2016). His remains were found in a rock shelter, like that shown on the right (B).

Galloway, A., & Wedel, V. L. (2014). Bones of the skull, the dentition and osseous structures of the throat. In V. L. Wedel & A. Galloway (Eds.), *Broken Bones* (Second ed., pp. 133–160). Springfield, Illinois: Charles C. Thomas.

Kremer, C., Racette, S., Dionne, C. A., & Sauvageau, A. (2008). Discrimination of falls and blows in blunt head trauma: Systematic study of the hat brim line rule in relation to skull fractures. *Journal of Forensic Sciences, 53*(3), 716–719. doi:https://doi.org/10.1111/j.1556-4029.2008.00725.x

Loe, L. (2009). Perimortem trauma. In S. Blau & D. H. Ubelaker (Eds.), *Handbook of Forensic Anthropology and Archaeology* (pp. 263–283). Walnut Creek, California: Left Coast Press.

Pfeiffer, S. (2016). An Exploration of Interpersonal Violence among Holocene Foragers of Southern Africa. *International Journal of Palaeopathology, 13*, 27–38. doi:https://doi.org/10.1016/j.ijpp.2016.01.001

Rao, N. K. (2003). *Textbook of Forensic Medicine and Toxicology*. New Delhi: Jaypee Brothers Medical Publishers Ltd.

Wedel, V. L., & Galloway, A. (Eds.). (2014). *Broken Bones: Anthropological Analysis of Blunt Force Trauma* (Second ed.). Springfield, Illinois: Charles C. Thomas.

instances were of other women and children. If these were instances of interpersonal violence, there was no simple analogy with ethnographically observed actions of biologically and culturally related groups. My first reaction was to shy away from the topic. I wanted to study humanity's successes, not its flaws. Ah, but one feature of research is that you must follow where the path leads. My motivation to understand the lives of Khoe-San hunter-gatherers led me to study and try to understand interpersonal violence.

Interpretive approaches to violent acts in the distant past

The way in which we interpret the human past is shaped in part by how we see and mentally organize the human present. There is an argument that this is an unavoidable fact of broadly framed narratives about human evolution (Landau, 1991). Whether a certain formula can be applied to interpreting all aspects of human evolution may be debatable (Godfrey & Cole, 1992; Padian, 2018), but writers who are based in hierarchical, patriarchal societies seem determined to see interpersonal violence as a fundamental and unavoidable feature of human nature. Books that explore human nature for a broad readership (Diamond, 2013; Pinker, 2011; Wilson, 2012) tend to portray the human past as unavoidably violent. For example, in a chapter titled "War as Humanity's Hereditary Curse," biologist E.O. Wilson writes "Archaeological sites are strewn with the evidence of mass conflict… Archaeologists have found burials of massacred people to be a commonplace (2012)".

Stanley Kubrick and Arthur C. Clark launched "2001: A Space Odyssey" with a scene in which an ape wields a newly discovered weapon (1968). The idea that aggression was a key to the success of our species was not new when that screenplay was written. The working assumption that violence was relatively common among distant human ancestors is sometimes labeled the "Hobbesian" perspective, after the English political philosopher Thomas Hobbes (Hames, 2019). In his 1651 poem *Leviathan* (Hobbes, 1946), he averred that it is human nature to act in one's own self-interest. Since self-preservation is the primary trait, conflict is inevitable. He described the natural state of humanity as a "warre of every man against every man," with life being "solitary, poor, nasty, brutish and short." In some modern writing, this ideology is described as being that of the Hawks.

Even back in Hobbes' day there were alternate views, although the most notable articulation of the opposing vision came about a century later. The proposition that warfare was rare or even absent outside of large-scale societies is sometimes labeled the "Rousseauian" view (Hames, 2019). Jean-Jacques Rousseau built his arguments on the proposition that human nature is a state of virtue, until it is corrupted by civilization. His position is encapsulated in his concept of the "noble savage" living in a state of nature (Rousseau, 1762). From this perspective, lethal violence was rare or absent in small-scale societies. As I approached the apparent evidence for interpersonal violence among southern African hunter-gatherers, I was definitely a Rousseauian. I was a Dove.

The skeletal evidence

Healed cranial trauma

It is not unusual to find a localized indentation on an adult skull from the Later Stone Age. There is no geographic or temporal pattern to where they are found (Fig. 11.1). It is in the nature of diploic cranial bone (see Box 11.1) to remain indented after healing from a blow that deformed the bone surface. Occasionally, there is enough context to reconstruct to some extent how the injury occurred. One example is provided in Fig. B11.1, in which perhaps the injury can be linked to rockfall. Another example is a small, sharp-edged fragment of stone still embedded in the skull of a man who lived about 3000 years ago, suggesting a nonfatal encounter with a weapon (Pfeiffer, 2016; Pfeiffer, Harrington, & Lombard, 2019). If the point had poison applied to its sides, as was the practice in preparation for hunting (Deacon, 1992), this was indeed a close call. If the man was the intended victim of violence, the attack was inept in both its target zone and its efficacy. The rest of the man's skeleton shows signs of osteoarthritis reflecting vigorous activity and a long life, but no other features that signal interpersonal violence.

A hunter-gatherer would have faced various hazards presented by animals, including venomous snakes, poisonous plants and carnivores. The skull of a woman who lived along the south coast about 1400 years ago shows multiple healed puncture wounds. Most notable are two large, smooth-edged holes, one on each side of the frontal bone. There is a depression near one of the holes. Taken together, the three wounds are consistent with three of the four canines of a predatory animal. The inter-canine space and lever arm of a brown hyena fit the cranial damage well (Pfeiffer, 2012, 2016). On the side where the temporal muscles would have been most badly damaged, the woman's jaws show dental destruction. There is less tooth wear one the more damaged side than on the other side. This is notable because

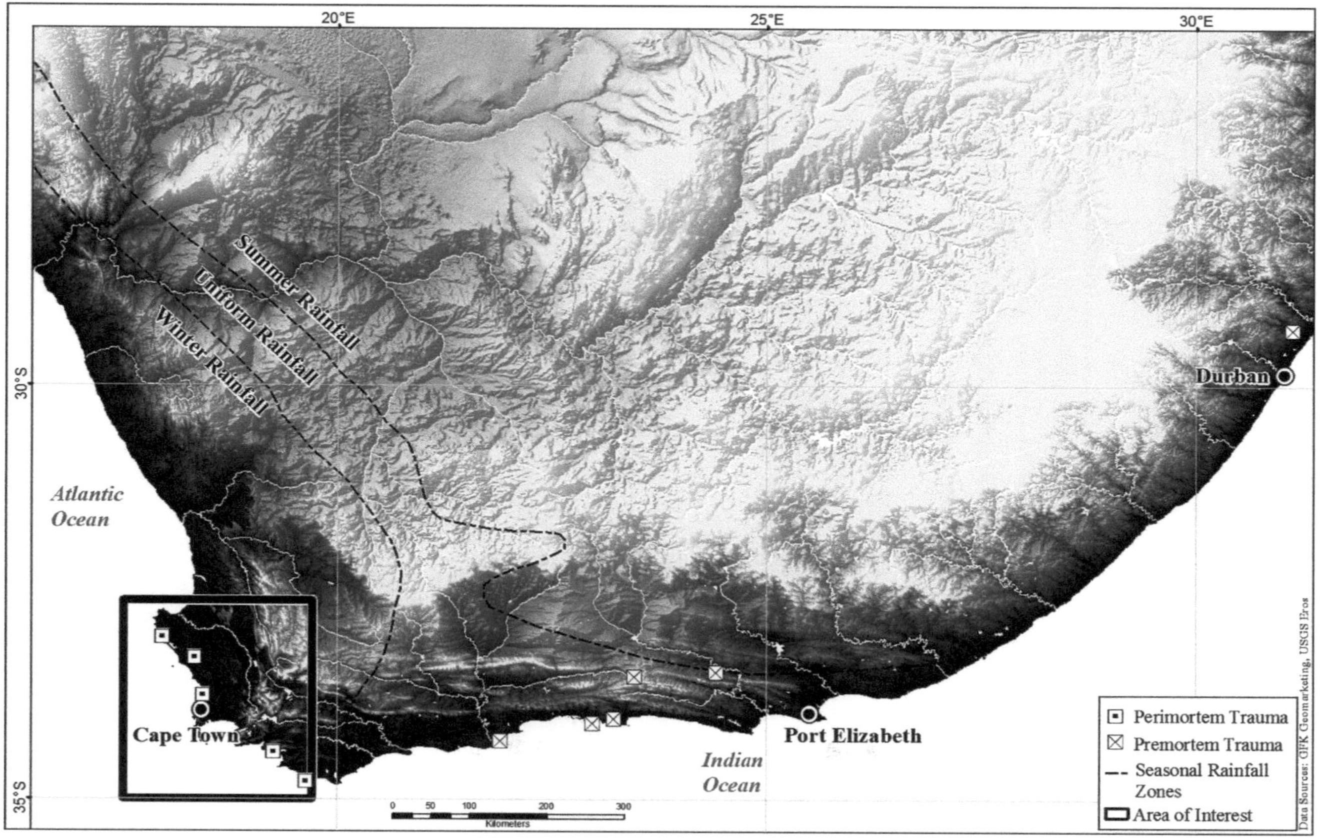

Fig. 11.1 Map of the southern African coastline. The skulls with premortem trauma described here are widely dispersed, while the skeletons with perimortem trauma come from a more limited area in the southwest. *(Source: Fig. 1, p. 29, Pfeiffer, S. (2016). An exploration of interpersonal violence among Holocene foragers of Southern Africa. International Journal of Palaeopathology, 13, 27–38. https://doi.org/10.1016/j.ijpp.2016.01.001.)*

it reinforces the evidence from the healed bone. The woman survived the attack and lived for quite a long time thereafter, although with compromised chewing and probably compromised speech. Modern evidence from such injuries suggests that mental function, too, could have been affected (Pfeiffer, 2016). How might one survive, when one's head was clenched in the jaws of a powerful predator? Surely a rescue by one or more people is a highly plausible guess.

Not all the healed cranial wounds can be attributed to accidents and dangers in the environment. Alan Morris and colleagues noted two indentations on the skull of a man who lived about 2400 years ago. Each is roughly oval in shape and has healed edges, although the original breakage is still clear. They look like they were caused by an object with a rounded profile—like a rod—brought down on his head with substantial force. The wounds are located toward the front, along what would have been the man's hair line. Their similar appearance and their shapes led the authors to conclude that the lesions were caused by interpersonal violence, from which the man recovered (Gibbon & Davies, 2020; Morris, Thackeray, & Thackeray, 1987).

These are examples of cranial trauma from which the victims recovered. Such instances reinforce what we would expect, namely that people encountered hazards in their daily lives. Sometimes their fellow humans were the source of the hazard, and sometimes their fellow humans may have saved them from further harm. Because of their disparate dates and unique features, they are not of much relevance to the question of whether organized interpersonal violence was a part of this hunter-gatherer society.

Unhealed cranial trauma

It would take quite a bit of force for a bone projectile point to penetrate through skin and muscle and lodge in the bone of a vertebra. The presence of three points in the spine of the woman buried at Quoin Point suggests an attack from close range on an unmoving victim. These features of close range, multiple "blows" delivered to a victim who seems an improbable combatant also apply to other discoveries.

Two other discoveries have these same features. The skeletons of a woman and a teen-aged girl were found, limbs in disarray, in a pit that had been closed with a large rock. Their skulls showed trauma consistent with the delivery of perforating and slashing blows, one gash showing an uneven breakage pattern consistent with a percussive blow from a flaked stone artifact. The woman's skull has an indentation to the left side of the skull consistent with blows to the right side while her head was resting against a stationary surface (Pfeiffer, van der Merwe, Parkington, & Yates, 1999). Again, there are multiple blows delivered at close range.

A third example is of three children, found buried together in a sand dune near the mouth of a river. A snapshot taken by the beachgoers who found the skeletons suggests that they were positioned in their grave with some care. The two older children (ages 6 and 12 years) were back-to-back, legs and arms folded in close, and the baby was at their feet. All three have cranial trauma. In this case, at least one of the weapons was a tube of round cross-section, like a digging stick. The skulls of the older children had been perforated multiple times (Pfeiffer & van der Merwe, 2004) (Fig. 11.2). The baby took a blow to the lower back of the head. We studied the long bone growth and the tooth development of the children. Their growth pattern had been normal. They had experienced only a few instances of growth disruption during the development of their tooth crowns (Harrington & Pfeiffer, 2016). They were not unusual in any biological respect. Once again, these victims appear to have been subjected to more physical violence than would have been needed, if the sole goal had been to kill them.

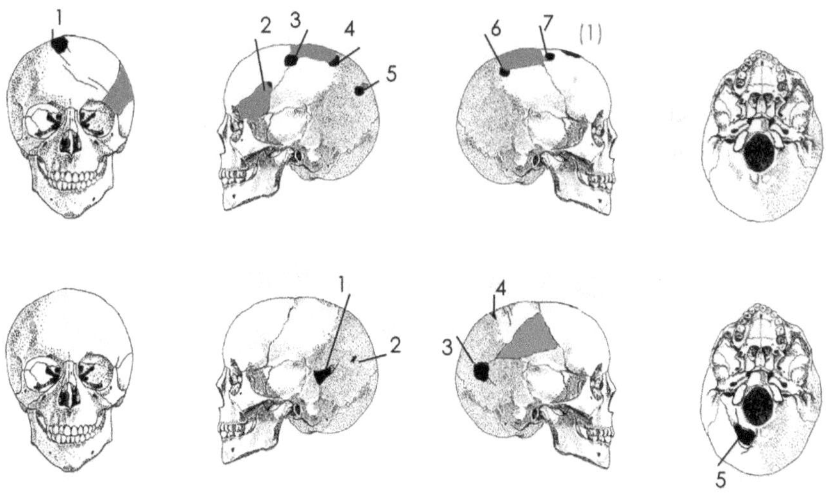

Fig. 11.2 Line drawings illustrating the penetrating wounds on the skulls of children buried in a dune at the mouth of the Modder River; the older child above, the younger below. Shaded areas indicate absent bone, numbers indicate perimortem wounds. *(Modified from Fig. 3, p. 61, Pfeiffer, S., & van der Merwe, N.J. (2004). Cranial injuries to Later Stone Age children from the Modder River Mouth, Southwestern Cape, South Africa. South African Archaeological Bulletin, 59(180), 59–65.)*

There are other examples of cranial trauma from which the victim did not recover. In two cases, Lagoon Beach and Faraoskop, skeletons with perimortem cranial trauma were found buried with other skeletons that are lacking that evidence. Circumstantially, it seems likely that their co-burial signals collective deaths from interpersonal violence. And then there are instances that are more difficult to interpret, where a skull from an undocumented context appears to show perimortem trauma, or where a skeleton is found in a grave with a bone point very near the body, such that it could have delivered poison that caused death.

What to make of it all?

As summarized in Table 11.1, the radiocarbon dates associated with most of these instances of unhealed trauma range from about 2500 to 1800 years ago. The latitudes and longitudes indicate that they've all been found within a relatively small, southwestern part of the fynbos region (Fig. 11.1). It corresponds to a coastal region extending about 100 miles north and south from the modern city of Cape Town. There are some differences among the group of cases. The Faraoskop group was buried within a rock shelter that may have been their home, rather than in coastal sands. Faraoskop is the only noncoastal discovery site. Not much has yet been published about the Faraoskop wounds. The Lagoon Beach group was buried in a section of coast that included other burials from about the same time—a possible cemetery (Pfeiffer et al., 2020). These two group burials are also a bit more recent than the other cases. It is not clear whether all the instances listed here represent the same cultural phenomenon, and of course new instances may be discovered.

Some broad observations may help to make sense of the evidence so far. Many ancestral Khoe-San skeletons have been discovered from the time period in which the suspicious deaths occurred. There may have been a bit of a population explosion, compared to prior and subsequent time periods (Fig. 9.2). Adult femur lengths generally trend toward shorter stature during this populous period (Fig. 10.1), but there is no decline in stature in the region in which the unhealed wounds are found. This distinguishes the region from the broader trend (Pfeiffer, 2016). This suggests that the region was the territory of a group that had distinctive practices with regard to resource exploitation. The victims do not have the demographic characteristics of combatants. That is, they are not adult men. This was not warfare, in that sense. None of the victims show evidence of defensive wounds to

Table 11.1 Skeletons associated with possible interpersonal violence from a region of southwestern South Africa (Fig. 11.1 map), dating from about 2500 to 1800 BP.

Site name (catalogue #)	Date BP (lab #)	Locale	Description	Citations
Quoin Point, Hermanus (UCT 317)	2220 ± 40 (Beta 241163)	34°78′S 19°63′E	Adult woman, newborn—peri-mortem bone points in spine	(Morris & Parkington, 1982)
Melkbosstrand (SAM-AP 6348 A,B)	2460 ± 50 (GX-23455)	33°73′S 18°43′E	Adult woman, adolescent girl-peri-mortem cranial trauma	(Pfeiffer et al., 1999)
Modder River (SAM-AP 6054 A,B,C)	2530 ± 60 (Pta-4151)	33°28′S 18°33′E	Three children, ages 1, 6, 12 years—peri-mortem cranial trauma	(Pfeiffer & van der Merwe, 2004)
Langklip, Hermanus (UCT 332)	2560 ± 40 (Beta 241164)	34°42′S 19°25′E	Adult woman—peri-mortem cranial trauma	(Morris, 2012; Sealy, 2010)
More recent dates				
Lagoon Beach (SAM-AP 6419–6424)	1840–1890 (4 dates)	33°53′S, 18°28′E	Four people buried together, one woman with evidence of powerful blow to back of head	(Pfeiffer, Sealy, Harrington, Loftus, & Maggs, 2020)
Faraoskop Rock Shelter (UCT 385–398)	2000–2100 (6 dates)	32°07′S, 18°36′E	Burial of twelve people, some with perimortem cranial trauma	(Parkington & Dlamini, 2015; Sealy, Patrick, Morris, & Alder, 1992)
No trauma, but reported as suspect because projectile points are present				
Saldanha Bay (SAM-AP 6372a)	2270 ± 40 (UGAMS 2802)	33°03′S 17°96′E	Adult woman—bone point found near cervical vertebrae	(Dewar, 2010)
Buffel's Bay, Knysna District (UCT 591)	2460 ± 40 (Beta 263,612)	34°08′S 22°97′E	Adult man–police recovery, including a bone point	(Pfeiffer, 2013)

the arms or hands. It is plausible that they were not conscious when they were attacked. Most of them sustained more than one wound. They were all members of a society that had extensive knowledge of neurotoxins, as used for hunting. A dose of neurotoxin could kill someone without leaving any skeletal evidence if the intent was simply to end someone's life. I think that the actions that led to the damage seen on these skeletons had more complex motivation.

Might there have been marauding invaders, targeting those least likely to successfully flee? Dated sheep bones from archeological sites indicate that the herding of domesticated animals had reached this region by about 2000 BP (Lander & Russell, 2018), a date quite a bit more recent than the onset of the interpersonal violence documented in these burials. There is considerable debate about whether the new mode of subsistence came through diffusion of the idea of pastoralism, or the arrival of new people. There is very little archeological evidence for pastoralist (Khoekhoe) pottery and occupational sites until much later (Sadr, 2008). During the period in which Khoekhoe pastoralists and Khoe-San foragers coexisted prior to the arrival of Europeans, there is no evidence of organized, inter-group violence. The evidence for marauding invaders 2500–1800 years ago just isn't there.

Warfare, or coalitionary violence, can be organized in various ways. "Socially organized armed combat between members of different territorial units" (Ember & Ember, 1992) is not the only framework. There can be feuds—a form of coalitionary violence within a community. In conflicts between communities or within a community, there can be "social substitutability" (Kelly, 2002) in which any member of a community can be a target of attack by virtue of their membership. In the case of ancestral Khoe-San people of a narrowly delimited time and place, a situation of feuding seems the most probable context for these foci of intense violence. From the scant evidence at hand, we cannot know what characteristics might have made certain women and children the targets of violence. The weapons seem to have come from the community's toolkit, yet they were applied in very uncharacteristic ways, like the use of a digging stick to injure a child. There are so many aspects of these wounds that seem uncharacteristic of the broader framework of ancestral Khoe-San life, I find myself leaning toward ritual explanations. That line of thought soon hits a dead end, for lack of firm evidence. The motivations of the executioners will remain a mystery.

For most millennia, throughout most of the land, there is no evidence of purposeful, fatal interpersonal violence among ancestral Khoe-San people. Its absence is the norm. As argued by Hames, the ability of hunter-gatherer bands to live peaceably with their neighbors reflects a uniquely human trait that deserves more attention from scholars (Hames, 2019). Nevertheless, I must admit that Hobbes had a good point. We humans have always been capable of harming one another. Our reasons can be diverse and idiosyncratic. The presence of interpersonal violence within one geographic region of southern Africa for a few hundred years is not analogous to any documented activities of historic or modern Khoe-San communities. It serves to illustrate a larger point: the past is not just a faint shadow of the present. Archaeology and bioarchaeology have the potential to discover transient human behaviors, which in turn reflect human nature. The details may not be discernable, but the broad patterns can help us to understand what it means to be human.

Bearing witness

I hope that this chapter and the previous chapters about southern African foragers have illustrated that osteobiographies can outline the people and events of past times and places. There are other stories that have been told about ancestral Khoe-San people, through the systematic study of several hundred individual burials, spread over a huge geographic expanse, each dated to a single end-of-life event. Through the collection of stories, we know more about the people who exploited the southern-most African landscape for thousands of years. The three chapters here tell a few stories, but far from all. The stories are not always in full agreement. On the one hand, studies of the crania and teeth show notable homogeneity through time and space (Irish, Black, Sealy, & Ackermann, 2014; Stynder, Ackermann, & Sealy, 2007). There is no evidence from those studies of significant barriers to the exchange of genes. On the other hand, in some locales, stark differences in dietary isotopes indicate that one community put up barriers to attractive food resources, keeping neighbors from sharing (Sealy, 2006). This probably represents territoriality. The similarities among skulls and teeth lead us to believe that strictly enforced territoriality, with barriers to mating, was rare. The topic of territoriality is one of many that could be explored further. As with human societies everywhere, there are stories at different levels, at different scales. And, just like with modern societies, there are more stories to be discovered.

The skeletons of past southern African foragers—the ancestors of the many modern ethnicities who trace their ancestry to Khoe-San peoples—provide opportunities for research approaches illustrating different priorities.

Research approaches to human remains from archeological contexts can focus on testing evolutionary hypotheses. The exploration of adult body size among Khoe-San ancestors, including the link between pelvic dimensions and linear growth, illustrates this kind of research. Bioarchaeological research can also inform our understanding of "anthropological generalizations." I'm using this term to refer to those overarching generalizations about human nature that form building blocks for our assumptions about humanity. They can focus on diet (what our ancestors ate must be good for us!), sex (who our ancestors found attractive, and what they did about it), warfare, and so on. The exploration of interpersonal violence among past foragers, determining its limited extent and its puzzling aspects, is an illustration of this kind of work.

Bioarchaeological research can also inform descendants about the past lives of their ancestors, providing stories about those times and places that are outside the scope of written documents. As I write this, the research on Khoe-San ancestors is not as broadly known among descendants as I think it should be. The postapartheid reclamation of indigenous identity is underway in southern Africa, but vehicles for communication with those who self-define as having Khoe-San ancestry are not obvious. I will keep an active watch on this front. Like all my colleagues in the Africanist archaeology community, I see the de-colonization of the African past as an important and challenging goal. It is my hope that Khoe-San descendants can be equipped to construct and expand their story line—one that is properly framed in time and space, in which their rich cultural heritage is supplemented by scientific approaches to the evidence.

References

Deacon, J. (1992). *Arrows as agents of belief amongst the /Xam bushmen*. Cape Town: South African Museum.

Dewar, G. (2010). Late Holocene burial cluster at Diaz Street Midden, Saldanha Bay, Western Cape, South Africa. *South African Archaeological Bulletin*, *65*(191), 26–34.

Diamond, J. (2013). *The world until yesterday: What can we learn from traditional societies?*. New York: Viking Penguin.

Ember, C. R., & Ember, M. (1992). Warfare, aggression, and resource problems: Cross-cultural codes. *Behavioral Science Research*, *26*, 169–226.

Gibbon, V. E., & Davies, B. (2020). Holocene Khoesan health: A biocultural analysis of cranial pathology and trauma. *International Journal of Osteoarchaeology*, *30*(3), 287–296. https://doi.org/10.1002/oa.2854.

Godfrey, L. R., & Cole, J. R. (1992). Narratives of human-evolution—Landau, M. *American Journal of Physical Anthropology*, *88*(1), 117–120. https://doi.org/10.1002/ajpa.1330880113.

Hames, R. (2019). Pacifying hunter-gatherers. *Human Nature—An Interdisciplinary Biosocial Perspective*, *30*(2), 155–175. https://doi.org/10.1007/s12110-019-09340-w.

Harrington, L., & Pfeiffer, S. (2016). Developmental variation in perikymata expression in co-interred child foragers. *American Journal of Physical Anthropology, 159*(S62), 168.
Hobbes, T. (1946). *Leviathan*. Oxford: Basil Blackwell.
Irish, J. D., Black, W., Sealy, J., & Ackermann, R. R. (2014). Questions of Khoesan continuity: Dental affinities among the indigenous Holocene peoples of South Africa. *American Journal of Physical Anthropology, 155*(1), 33–44.
Kelly, R. C. (2002). *Warless societies and the origin of war*. Ann Arbor, Michigan: University of Michigan Press.
Landau, M. (1991). *Narratives of human evolution*. New Haven, CT: Yale University Press.
Lander, F., & Russell, T. (2018). The archaeological evidence for the appearance of pastoralism and farming in Southern Africa. *PLoS One, 13*. https://doi.org/10.1371/journal.pone.0198941.
Lee, R. B. (1979). *The !Kung San: Men, women and work in a foraging society*. Cambridge: Cambridge University Press.
Marshall, E. M. (1958). *The harmless people*. New York: Knopf.
Morris, A. G. (1992). *A master catalogue: Holocene human skeletons from South Africa*. Johannesburg, South Africa: Witwatersrand University Press.
Morris, A. G. (2012). Trauma and violence in the Later Stone Age of Southern Africa. *South African Medical Journal, 100*(6), 3–4.
Morris, A. G., & Parkington, J. (1982). Prehistoric homicide: A case of violent death on the Cape south coast, South Africa. *South African Journal of Science, 78*, 167–169.
Morris, A. G., Thackeray, A. I., & Thackeray, J. F. (1987). Late Holocene human skeletal material from Snuifklip, near Vleesbaai, southern Cape. *South African Archaeological Bulletin, 42*, 153–160.
Padian, K. (2018). Narrative and "anti-narrative" in science: How scientists tell stories, and don't. *Integrative and Comparative Biology, 58*(6), 1224–1234. https://doi.org/10.1093/icb/icy038.
Parkington, J. E., & Dlamini, N. (2015). *First people: Ancestors of the San*. Cape Town, South Africa: Krakadouw Trust.
Pfeiffer, S. (2012). Two disparate instances of healed cranial trauma from the Later Stone Age of South Africa. *South African Archaeological Bulletin, 67*(196), 256–261.
Pfeiffer, S. (2013). Population dynamics in the Southern African Holocene: Human burials from the West Coast. In A. Jerardino, D. Braun, & A. Malan (Eds.), *Vol. 84. The archaeology of the West Coast of South Africa: Cambridge monographs in African archaeology* (pp. 143–154). Oxford: Archaeopress.
Pfeiffer, S. (2016). An exploration of interpersonal violence among Holocene foragers of Southern Africa. *International Journal of Palaeopathology, 13*, 27–38. https://doi.org/10.1016/j.ijpp.2016.01.001.
Pfeiffer, S., Harrington, L., & Lombard, M. (2019). The people behind the samples: Biographical features of past hunter-gatherers from KwaZulu-Natal who yielded aDNA. *International Journal of Palaeopathology, 24*, 158–164. https://doi.org/10.1016/j.ijpp.2018.10.008.
Pfeiffer, S., Sealy, J., Harrington, L., Loftus, E., & Maggs, T. (2020). A Late Holocene community burial area: Evidence of diverse mortuary practices in the Western Cape, South Africa. *PLoS One, 15*(4). https://doi.org/10.1371/journal.pone.0230391. e0230391.
Pfeiffer, S., & van der Merwe, N. J. (2004). Cranial injuries to Later Stone Age children from the Modder River Mouth, Southwestern Cape, South Africa. *South African Archaeological Bulletin, 59*(180), 59–65.
Pfeiffer, S., van der Merwe, N. J., Parkington, J. E., & Yates, R. (1999). Violent human death in the past: A case from the Western Cape. *South African Journal of Science, 95*, 137–140.
Pinker, S. (2011). *The better angels of our nature: Why violence has declined*. New York: Viking.
Rousseau, J.-J. (1762). *The social contract, or principles of political right (Du contrat social)*.

Sadr, K. (2008). Invisible herders? The archaeology of Khoekhoe pastoralists. *Southern African Humanities, 20,* 179–203.
Sealy, J. C. (2006). Diet, mobility and settlement pattern among Holocene hunter-gatherers in southernmost Africa. *Current Anthropology, 47*(4), 569–595.
Sealy, J. C. (2010). Isotopic evidence for the antiquity of cattle-based pastoralism in southernmost Africa. *Journal of African Archaeology, 8*(1), 65–81.
Sealy, J. C., Patrick, M. K., Morris, A. G., & Alder, D. (1992). Diet and dental caries among Later Stone Age inhabitants of the Cape Province, South Africa. *American Journal of Physical Anthropology, 88,* 123–134.
Skotnes, P. (2007). *Claim to the country: The archive of Wilhelm Bleek and Lucy Lloyd.* Johannesburg, South Africa and Athens, Ohio: Jacana Media and Ohio University Press.
Stynder, D. D., Ackermann, R. R., & Sealy, J. C. (2007). Early to mid-Holocene South African Later Stone Age human crania exhibit a distinctly Khoesan morphological pattern. *South African Journal of Science, 103,* 349–352.
Wilson, E. O. (2012). *The social conquest of earth.* New York: W.W. Norton.

CHAPTER 12
Repatriation in an African context

Prologue

It is presumptuous for anyone to attempt a brief summary of any activity as it has occurred across an entire continent. This overview will include only a few examples of activities, from those regions where I have some familiarity with the actors and institutions. Despite this limited scope, I hope to show that there is a pattern to how repatriation fits into a broader theme of cultural restitution. In the context of this book, the way repatriation is approached in Africa has parallels with the Canadian example described in Chapter 6.

When it comes to past collection of human skeletons, a dominant theme across the continent of Africa is the co-occurrence of colonialism and race science. Whether it was the Belgians in the Congo, the Germans in the Namib, the British in South Africa and Kenya, the Portuguese in Angola and Mozambique, the French in West Africa, Spaniards in Morocco, or other colonial power elsewhere, the Scramble for Africa (1876–1912) (Pakenham, 1991, p. 300) coincided with the era in which race science was booming (Fig. 12.1). In the 19th and early 20th centuries, it was very widely accepted that races were biologically real, distinct, and could be ranked. The dominance of the colonial forces, combined with their assumption of racial superiority, led to the procurement and export of the skeletons of Africans to institutions where their novel biological features could be studied.

There are often documents describing where and when the skeletons were exported. Thanks to the relatively recent dates and the transactional precision of the collectors, the curated entries, making up what are referred to as legacy collections, can often be linked to historical events. As we hurtle along in the 21st century, you might expect that the curated African skeletal collections held in Europe and North America would be arriving at African ports of entry in shipping containers, being delivered to the governments of the now independent African countries from whence they came. However, there are impediments to their return. The world does not stand still, and circumstances in the communities from which bodies were taken have often

Osteobiographies
https://doi.org/10.1016/B978-0-12-823880-6.00013-6

Fig. 12.1 The Scramble for Africa. (Credit: James Gillray, "The Plumb-pudding in danger," 1805.)

changed dramatically. The identities of the skeletons are often incompletely known. The recipient communities may be ill-prepared to face the expense and complexity associated with getting them back. For these reasons and more, solutions have taken different forms. In some instances, solutions are not yet apparent.

Unnamed persons

African authority over European holdings

The era of Belgian rule in Central Africa lasted until 1960. Especially during the reign of King Leopold II (1865–1909), treatment of people in the Belgian Congo was notoriously harsh. The regime relied on forced labor to cultivate and trade rubber, ivory and minerals (Hochschild, 1998). The Royal Museum for Central Africa (RMCA), in Tervuren, Belgium, was built for the 1897 World Exhibition to showcase King Leopold II's loot from Congo. Human skeletons were part of such collections, since the racial characteristics of the various tribal groups were to be documented, as part of the exotic Congo world. A study has been undertaken at the Belgian government's request to identify and itemize the human remains. The spoils of conflict are believed to include hundreds of skeletons, as well as skulls from the heads of rebellious African leaders. Most of the human remains are said to have been transferred from the RMCA to the Royal Institute of Natural Sciences of Belgium (Bouffioux, 2020).

It is no surprise that other Belgian institutions dating from that era also hold materials of tainted provenience. The collection of about a dozen colonial-era skulls held by the Free University of Brussels (ULB) is small. Nevertheless, its existence became a focus for the ULB community. In 2019, the ULB held a colloquium entitled "From shadow to light—For a policy for the management of colonial collections of human remains in universities." Various icons and documents within the institution were incorporated into the exploration. No request for return of the remains had been initiated by any Congolese institution, but the ULB set out to make a plan for the skulls' return. In August 2020, an agreement was signed between ULB and the University of Lubumbashi (UNILU). It transfers conservation rights to UNILU, an institution that ULB was instrumental in founding at the end of the colonial era. The two institutions have close ties. The plan is that the skulls will stay at ULB for the next five years, during which the UNILU has the authority regarding any new research. It is the shared intention of the two universities to apply methods like CT scans and 3D reconstruction, which rely on technology available in Europe but not at UNILU. At the end of five years, the skulls will be returned to Congo unless UNILU requests a delay. In a maximum of eight years, the skulls will be sent home.

The process followed in this case has parallels with that of the Huron-Wendat Nation of Canada, described in Chapter 6 The ULB knew the general circumstances through which the skulls arrived, and they determined that they could not learn the specific identities of the individuals. The return would need to be to a region, not a specific family. Within the broad rubric of "Congo," a locus of leadership was needed. There are dozens of universities in the Democratic Republic of Congo, so this step might have been quite a challenge. The ULB reached out to the institution at Lubumbashi, and they made a plan. The first formal step of that plan was the transfer of authority. The receiving entity now has authority over the skulls and will determine what happens to them from now on. Perhaps this case can provide a template for the disposition of the larger, more diverse, colonial-era collection of skeletons and associated material items from the Congo.

Nama and Herero genocide

There are many examples of conflicts in which the victors treat their slain enemies with cruelty. The sword wounds delivered to the carcass of English King Richard III were used to positively identify his skeleton (Buckley et al., 2013). A different kind of contempt was directed toward the dead during 1904–08, as the Germans controlling what was then South West Africa (now Namibia) attempted to exterminate the Nama and Herero peoples. Many San were also drawn in. The catalyst was a Nama rebellion against German colonial rule, started in January 1904. Many thousands of people died, as people of all ages were driven into the Namib desert and imprisoned in concentration camps. Drawn to this ongoing genocide came Eugen Fischer: a German professor of medicine, anthropology and eugenics (and a member of the Nazi Party). In 1906, in the context of his Namibian "field research," he collected skeletons and sent them back to Germany. The thrust of his work was firmly linked to racial purity and the perils of miscegenation. In 2004, at the 100th anniversary of the start of the genocide, Germany expressed an official apology. That act animated a search to find and return human remains that could be linked to the genocide.

To contemporary Nama, Herero and San peoples, the bones archived in Germany are symbolic of the deaths of what some estimate to have exceeded 100,000 innocent people. With German cooperation, many crania have been returned. In 2011, twenty arrived from the Berlin University Hospital Charity Museum; in 2014, 35 arrived from University of Freiburg; in 2018, 28 arrived from various private collections in Berlin. Most of them have very little accompanying information (J. Sylvester, pers. com. 03/2019). An investigation by the University of Freiburg, where Fischer was curator, concluded that he brought approximately 200 skulls from Africa (Freiburg, 2014). More may be sent to Namibia, as their association with unethical acts becomes known to those who possess them. Iziko Museums in Cape Town have identified a large number in their collection (Black & McCavitt, 2020:125).

As I write this, the returned skulls are being held at the National Museum of Namibia in Windhoek. Their final disposition remains unresolved. Living people who identify with the genocide now reside in many countries, oftentimes because their ancestors fled the mayhem. It would be daunting to attempt personal identification of each skull. Genomic information might be valuable, but who would give permission for the destructive test? If they are to be buried, where should they be interred? These few skulls are a physical reminder of a bleak era in southern African history. Those who now hold them in trust for the affected communities have a big responsibility.

Culturally significant Mapungubwe burials

The archeological site of Mapungubwe is central to many strands of the past for Black South Africans. Clustered around an impressive massif near the confluence of the Limpopo and Shashe Rivers, Mapungubwe was the center of an 11th to 13th century C.E. stratified, urban society with far-reaching influence. It is a precursor to the ancient city and kingdom known as Great Zimbabwe. That site is a central symbol for modern Zimbabweans, South Africa's neighbors on the other side of the Limpopo. Extensive excavations of Mapungubwe were undertaken by archeologists of a colonialist persuasion. Their explanatory framework for the obvious cultural richness and complexity relied on imagined foreign influence. A site adjoining Mapungubwe was named "K2" because its deep occupational deposits reminded the excavator of tell-like sites of North Africa (Mitchell, 2002:300). It could not have been Black Africans who achieved all this! Contemporary interpretations have identified clear linkages between the archeological evidence and the Indigenous cultures of the region. Black South Africans rightly see it as their heritage.

A centerpiece of the discoveries was a cemetery on top of the massif where members of the ruling lineage were interred with golden jewelry. When archeologists from University of Pretoria disinterred the royal burials in the 1930s, three of 27 were found to have gold inclusions. The presence of glass beads, copper and gold at Mapungubwe Hill indicates integration into a trading network. This site, far from the coast, was linked to the coast of what is now Mozambique, from which trade was linked by ship to the coastal community at Kilwa Kisiwani in what is now Tanzania, almost 2000 miles away. The gold objects include a scepter, a bowl (which perhaps was a crown), beads and animal figurines. The golden animals include a crocodile, an ox, a wild cat and fragments of others, but best preserved and most remarkable is a golden rhinoceros.

The Golden Rhino of Mapungubwe is today an important South African cultural identifier (Fig. 12.2). It was designated a National Treasure in 1999 (Tiley-Nel, 2016). In 2002 the ruling party created the Order of Mapungubwe, the highest honor in South Africa. The golden rhino is a symbol of a southern African Naissance (Leslie and Maggs, 2000), well prior to the arrival of Europeans. It is in this context that support grew for the idea of reburying the royal skeletons and those from archeological sites closely associated with Mapungubwe. The Mapungubwe Cultural Landscape is now a national park and a World Heritage Site. There, after

Fig. 12.2 The Golden rhinoceros of Mapungubwe; about 11 cm long, made of foiled 24-carat gold.

broad community consultation and extensive preparation, the University of Pretoria reburied human remains from seven related sites ($N = 143$) in 2007 (Nienaber, Keough, Steyn, & Meiring, 2008).

Remains of enslaved Africans who died far from home

The skeletal collections of 19th century "race science" were built not just of bones shipped from abroad. They could also take advantage of the deaths of enslaved persons. The African diaspora associated with slavery transported people from diverse regions of Africa to countries in Europe and the Americas. The skeletal remains of a tiny fraction of those enslaved persons were catalogued in collections after their deaths. Particularly well documented are the crania of Africans in the Collection of Samuel George Morton (1799–1851), now housed at the University of Pennsylvania Museum of Anthropology. The crania have been CT scanned, forming an online archive that attracts scholars from many fields. Much has been written about this collection (Renschler & Monge, 2013). Of the original 867 crania amassed by Morton, 82 are categorized as "Native African." Among the Native Africans are a number (between 51 and 54, some documents being unclear) sent to Morton in 1840 by a physician in Havana, Cuba. They were described by the sender as having died soon after their arrival in Cuba. They had been buried in Cuba. If the well documented 51 African-via-Cuba crania were to be repatriated, surely reburying them in Cuba would be wrong-headed. Several have filed anterior teeth or evidence of anterior tooth ablation, which is another confirmation of their African origin. Dental modifications of these sorts were common across much of Africa during the period of the slave trade (Friedling & Morris, 2005; Goose, 1963). The documents associated with some of the Morton "Native African" crania provide fragmentary information about their origin, but certainly not enough to make confident attributions of their national birth origin.

There are other collections at other institutions. Harvard University has identified 15 possibly enslaved persons in their collections (Bolotnikova, 2021). The broadly shared goal of the race science era to define "different varieties of man" (Stanton, 1960) motivated the trade and aggregation of African crania, with those of enslaved Africans suiting the purpose. If repatriation is to be a vehicle for restorative justice, what should be done in these cases? Genomic and palaeogenomic research is beginning to describe the diversity of African communities (Wang et al., 2020). One approach, in principle, would be to match the aDNA of each cranium to a putative modern community. The cost and complexity of such an approach would be substantial. Another approach would be to make common cause with existing slave trade memorials in Africa, like that of the embarkation site of Gorée Island, Senegal. The skulls of enslaved persons could at least be buried on their continent of birth. Another approach would be to consider collected African crania as part of discussions about memorializing enslaved people in the countries where they were enslaved and died. Enslaved persons born in Africa have been included at the African Burial Ground National Monument, New York, NY. The topic may become integrated into the idea of federal legislative protection of the graves of African Americans, similar to NAGPRA, which has been suggested (Dunnavant, Justinvil, & Colwell, 2021). Inclusion of the skeletal remains of humans who had never been enslaved, like many of the crania in legacy collections, would be an extension of scope for such memorials. While the focus here is on Africa, it is worth mentioning that legacy collections include many "orphan crania" from locales around the world. Their existence is a reminder that repatriation initiatives can present a significant conundrum about where, and to whom, bones are to be sent.

Named persons

Skeletons in the cupboard

April 27 is celebrated as Freedom Day in South Africa. It marks the date of the country's first multiracial parliamentary elections in 1994. Subsequently, during Nelson Mandela's presidency, South Africans initiated sweeping initiatives aimed at affirmation of ethnic diversity and reconciliation for past inequities. A slim volume representing one of these initiatives is the work of Martin Legassick and Ciraj Rassool, *Skeletons in the Cupboard: South African museums and the trade in human remains 1907–1917* (Legassick and Rassool, 2000). The facts of grave robbing and sale of human skeletons were not newly discovered. They had been hiding in plain sight. The stories of Dickensian characters like Dr. Pöch and his shovel-wielding assistant Mr. Mehnarto who fulfilled orders for "specimens," are part of the region's history. The central message, forcefully expressed by Legassick and Rassool, was that the curation by public institutions of skeletons of known people, taken without permission and against the will of their families, is not acceptable. They argued that these bodies should be repatriated.

The volume includes essays of response from curators of four of the country's museums as well as the South African Museums Association. The essays are courteous and somewhat cautious. The writers could see that their jobs were about to get more difficult, but they were on board. In a clear a sign of support, the book was published by the South African Museum (now called Iziko Museums). In concert with the book's release, the museums of South Africa and neighboring countries undertook to review their holdings and identify any catalogue entries that were "unethically collected." Study access to these materials would henceforth be banned, as the institutions considered their next moves. Many of the fresh bodies had been sent to Europe. Hence, the idea of identifying and banning access to unethically collected holdings spread to include some European institutions. South African researchers conducted on-site reviews of international collections, which helped to trace the whereabouts of the exported human remains (Steyn, Morris, Mosothwane, Nienaber, & Maat, 2013) (see also Chapter 13).

The next step envisioned by Legassick and Rassool, that of the materials' return to the community and reburial, has proved to be more challenging. Who, precisely, are these people? If there are names in the catalogue, will those names appear in historic records, unabiguously? If there is no name, can historic and scientific evidence provide productive clues? If an identity is established, which of their many relatives should be designated as descendants? Should this be based solely on biology, or does language and ethnicity carry weight? What scientific tools should be applied, in this exploration? If undisputed descendants can be identified, who is responsible for the costs associated with fulfilling those descendants' wishes? For example, who buys the burial plot? Some examples of repatriations that have been completed in recent years, and that are underway now, will illustrate the complexity of the matter. They will also illustrate the sustained commitment of many people, especially the curators and scientists responsible for the remains.

Sara Baartman: D. 1815, buried 2002

The petite woman of Khoe descent was born in the eastern part of the Cape Colony of South Africa in the late 1700s. Having become a domestic servant in Cape Town, she was taken to London by an English doctor in 1815. There, she was exhibited as "the Hottentot Venus." She became well known, not as a person but as a strange, exotic specimen. Her nakedness, tightly coiled ("peppercorn") hair, large buttocks (steatopygia) and prominent areolae aroused the Victorians. The physician sold her to an animal trainer who exhibited her in Paris. There, the famous comparative anatomist Baron Georges Cuvier took an active interest. He had drawings made of her in life, and he arranged to procure her body when she died in 1815, purportedly at age 26. Her brain, sex organs and skeleton were accessioned to the Musée de l'Homme, and there she stayed for over a century (Fauvelle-Aymar, 2002).

Based on what can be learned from documents of the time, Baartman's brief life was one of servitude and blatant exploitation. Her plight became a symbol for the British anti-slavery movement. She was also seen by scientists like Cuvier as a potential link to pre-human species.

(Crais & Scully, 2009). Her story is familiar to South Africans. Surely if the new, post-1994 South Africa were to take on a cause of redress, the repatriation of Sara Baartman would be an obvious choice. From 1997 to 2002, there was a sorting out of which group most appropriately represented her natal community (Morris, 2011:197–206). In early 2002, Sara Baartman was de-accessioned and sent home. She was given a Christian burial at a site in her home province on Women's Day (August 9).

Prior to that date, her skeleton was briefly available for study at the University of Cape Town. Alan Morris noted some features that are consistent with historic accounts and others that contradict it. Healed fractures of the face and forearm are consistent with her difficult circumstances. Her lower leg bones showed abnormal changes of the kind associated with venereal disease, which is thought to have been the cause of death. The maturation of her skeleton was consistent with an age at death somewhat older than 26 years, affirming recent historians' surmise (Crais & Scully, 2009). Although she is thought to have given birth two or three times, her pelvic bones do not show the bone changes that sometimes follow. She was not as short as Cuvier said she was. In 1815, Cuvier measured her to be 4′6″ (137.2 cm). Based on the length of her femora, Morris concluded that she was about 10 cm taller than that (Morris, 2011:201–202).

"El Negro of Banyoles": D.1830, reburied 2000

This is an account of a man whose given name is unknown, but who was known by a variety of titles during the subsequent 150 years. On a day in 1830, a young man was buried somewhere near the mountain town of Tulbagh, about 100 km inland from Cape Town. His given name and his ethnicity remain unknown. Although the region where he died is far to the south, it is possible that he had been born in what is now Botswana. The import to the Cape Colony of slaves from overseas had been banned in 1809. Raiding of countries just beyond the northern border for workers had become common, as documented by missionaries of the time. The Batswana were one of the targeted groups.

Within a day of his burial, this particular young man was dis-interred by the Verreaux brothers. These enterprising European taxidermists were in the region to collect specimens of African animals. They retained the man's skin, his skull and a few bones, discarding the rest. They prepared his body as a trophy for display and they shipped him off to Europe. He was first publicly displayed at Maison Verreaux in Paris. His label was "Le Betjouana" and he carried weapons that would have been typical of 19th century Tswana warriors. He remained there until sometime in the 1880s, when he was sold to a Spaniard, Francisco Darder. He was displayed in Barcelona as "El Bechuana," then he was moved to Banyoles, Spain, where he was labeled as "a Bushman from the Kalahari" in the Darder Natural History Museum. The sobriquet "El Negro" referred to his very dark skin, which was the product of shoe polish, periodically applied (Westerman, 2019).

This sorry history can be verified through photographs and public accounts (Fig. 12.3). The display of a stuffed human was a cause of concern to only a few (Westerman, 2019), until his image was included in tourist brochures prepared for the 1992 Barcelona Olympics. Outrage was expressed by people from diverse quarters. A Pan-African boycott of the games was considered. In response, Spanish authorities agreed to take up the issue. Despite the intensity of this brief cry of alarm, the man remained on display until 1998. This was when the Organization of African Unity requested that Botswana receive the body for reburial on behalf of the continent.

The Botswana government received his remains in early 2000 and were surprised to find that only the skull and "a few bones" had been sent. Spain had reported that a DNA sample had been taken previously (from the skin, perhaps?) but nothing came of that. They said that an autopsy had been done in 1995, concluding quite remarkably that the man had been 27 years

Fig. 12.3 In 1888, this image of the taxidermied corpse of El Negro appeared in a catalogue of a Barcelona museum, where Francisco Darder was curator. (Wikipedia, Negro_of_Banyoles; accessed Nov 16, 2020.)

old and had probably died of pneumonia! It seems improbable that a skull which at that point was still encased in a fake carcass could provide such precise information. The skull of El Negro was displayed at the State House in Gaborone, where hundreds are said to have viewed it. The remains were then buried in a public park in the capital city, accompanied by a commemorative plaque (Ramsay, 2019).

Klaas and Trooi Pienaar: D. 1909, reburied 2012

The prior examples of named persons who have been repatriated have taken the story line rather far from the argument made by Legasick and Rassool, that curated skeletons should be a focus of attention and redress. There have been some successes in this domain. In August 2012, two Northern Cape farmworkers were reburied at a site near that were they had been buried the first time, in 1909. A married couple, Klaas Pienaar and his wife Trooi, had died of malarial fever within a month of each other. They were in their 60's or 70's, and they had four surviving children at the time of their deaths. As was the custom, they were buried on the farmland, near the town of Kuruman. About four months later, they were dug up by the professional grave robber Mehnarto, on the instructions of Rudolf Pöch, Professor of Anthropology at the University ofVienna. Pöch had ordered specimens for his scientific study in Vienna, and Mehnarto found the Kuruman region very suitable for his assignment. The dry soil and hot temperatures deterred natural post-mortem deterioration, and the farm workers were Khoe-San people: the focus of Pöch's interest.

Mehnarto and his assistants are said to have wrapped the Pienaars in linen, cut their legs at the knees to fit, and encased them in salt inside a large barrel. The barrel was taken by wagon to Cape Town, where it joined stolen rock engravings and many other skeletons. The shipment was then exported (illegally) to Vienna. They became part of a collection (Rassool, 2015). How were they reclaimed and reburied? For anyone who suspects that the life of a scholar is boring, this story is for you. There can be some dramatic moments. Quoting the President of South Africa, from his speech at the reburial:

> *"In Vienna in May 2008, at a conference held at the Natural History Museum to celebrate Pöch who had been regarded as an accomplished scholar, our Professors Legassick and Rassool outlined the brutal details of how Pöch's human remains collection had been acquired. We are pleased that subsequent to that presentation and engagements, the Natural History Museum of Austria committed itself to return the remains of Mr Klaas and Mr Trooi Pienaar to South Africa. We are happy, too, that the Austrian Minister of Science then authorized the formal removal of the remains of Klaas and Trooi Pienaar from the collection of the Austrian Academy of Science. We take this opportunity to thank the Austrian government for taking this step and for working together with us to bring finality to this painful matter... Through today's reburial, their humanity is being reaffirmed and their dignity is being restored irreversibly."*
>
> ***(Zuma, 2016).***

Activities underway

Institutions throughout the African continent have taken steps to include input from descendant communities. Boards of oversight have been re-structured in several cases.

Progress toward a coordinated approach for dealing with unethically collected human remains is the goal of the Southern African Human Remains Management Project (SAHRMP). The group's goals are to "rehumanize" these individuals and to open a dialogue with indigenous populations, with an end goal of returning them (Black & McCavitt, 2020). International workshops in 2017, 2018 and 2019 were organized around coordinating approaches to shared challenges of identifying unethically collected remains, identifying descendants, consulting in productive ways, and finding funds for this work. Costs are linked not just to skilled staffing but also to extensive travel and translation.

The initiative by the University of Cape Town (UCT) is an example of both the challenges and the satisfactions of the work. When she reviewed the archived human skeletons in 2017, Prof. Victoria Gibbon at the UCT Faculty of Health Sciences noted one group of catalogued remains that had been unethically collected from near Sutherland, a small town in the Northern Cape, about 350 km north of Cape Town. In the 1920s, a medical school student had provided the anatomy professor with nine Khoe-San skeletons. The student had exhumed them from his family farm. The skeletons belonged to named persons who, like the Pienaars, had been buried on the farm where they were indentured. With two family names and a geographic origin, Gibbon hoped that a repatriation could be readily arranged.

With the support of UCT, Gibbon and a small team traveled to Sutherland and met with the community, to explain that their relatives' remains were at the University. How did they wish to proceed? The community was shocked and dismayed. They had many questions. From that point onward, the community directed what would be done with these individuals and their records. The Sutherland families sought to find out as much as they could about their relatives, and how they came to be at UCT. It was Gibbon's job to explain the options in a nondirecting way. Gibbon put together a team of local and international researchers. An archeologist identified the spot from which the people had been removed, and a historian provided the social historical context. Osteobiographies of each person were completed, describing their sex, age-at-death, stature, and any trauma or pathological features. Tiny samples of dental tissue were analyzed for their carbon and nitrogen isotope ratios. Samples for aDNA were analyzed by a laboratory at the Max Planck Institute for the Science of Human History in Germany, and a PhD student at Liverpool John Moores University in the U.K. created facial reconstructions. After national news agencies reported the scientific results to the public, UCT was approached by a family in the Eastern Cape who share one of the family surnames. Based on the predominance of that surname in their region, they claimed that one deceased person had originated there, rather than in the Northern Cape. The stable isotope results confirmed that this person had spent their childhood in the arid north of the country, not the more well-watered east.

In October 2019, the researchers traveled to Sutherland, to present their results to the community. Press releases from UCT and independent media reports describe a meaningful, dramatic event in which everyone involved sensed the importance of what they had accomplished (Cloete & Shaw, 2019; M-Net, 2019; Swingler, 2019; Wild, 2019). The transfer of responsibility had taken place, a collaborative partnership of scientists and community members had been forged, and new knowledge about past lives had been conveyed to people for whom that knowledge holds great value. As I write this, the re-interment has not yet taken place, delayed by disruptive factors like the global COVID-19 pandemic.

Challenges met

These brief stories share several themes. Why is repatriation such a challenge? Perhaps a charitable perspective is that people are set in their ways, and repatriation disrupts their routine. Or maybe repatriation requests are analogous to other acts of decolonization, generating initial feelings of defensiveness and uncertainty about how to respond. The examples here show that repatriation takes bold action followed by sustained determination and access to resources. Even when there is historical documentation and a general sense that a wrong was done, motivation to work toward redress may have lower priority than other pressing matters. Multiple collective entities and levels of governance may be involved. The human tissue that still exists is often much less extensive than what the petitioners anticipate. This reinforces their suspicion that multiple acts of disrespect occurred, which in turn sours the discussions. Once the remains are received, the decisions about if, where and how to bury or rebury them can be difficult.

The African context illustrates in vivid detail how the colonial era and the concurrent dogma of race science imposed indignities that are now broadly deemed to have been unethical. It is unrealistic to expect the indigenous communities to bear the full burden of initiatives for redress. Many are still re-formulating their identities after decades of apartheid-era suppression. Economic and educational disadvantages need attention. For example, representative organizations have established benefit-sharing agreements that recognize traditional knowledge about natural plant products, including Hoodia and rooibos (Schroeder, Chennells, Louw, Snyders, & Hodges, 2020). For the communities, repatriation is one of many competing priorities.

It is gratifying to see scholars from diverse backgrounds supporting indigenous groups, taking on much of the burden to find and send home the unethically collected bones. It is not enough to just describe what happened many decades ago, although this is a crucial first step. To the extent that they have the power to do so, my colleagues are acting to redress the past.

They share with society the goal of restorative justice. The proactive support of organizations like the International Council of Museums (ICOM), the Commonwealth Association of Museums (CAM), and individual museums across Africa will continue to be very important. Whatever the geographic and cultural space, making common cause with descendant communities is a challenge for scholars and scientists. This includes helping the community understand their options, then deferring to their decisions. There are barriers of language, jargon, habits and priorities. Nevertheless, momentum is growing. The "transferable insights" identified by the authors of the 2020 Rooibos Benefit Sharing Agreement can also be applied to acts of repatriation. For success, actors need to apply "patience; incrementalism; honesty; trust; genuine dialogue; strong legal support; a shared recognition that a fair, win-win deal is possible; government leadership; and unity among indigenous peoples" (Schroeder et al., 2020:285).

References

Black, W., & McCavitt, K. (2020). The Southern African human remains management project: Making (p)reparations in year one. In C. H. Meloche, L. Spake, & K. L. Nichols (Eds.), *Working with and for ancestors: Collaboration in the care and study of ancestral remains* (pp. 115–126). London, UK: Routledge.

Bolotnikova, M. N. (2021), Peabody museum discovers possible slave remains in its collections. Harvard Magazine, 1.28.21, Retrieved from https://harvardmagazine.com/2021/01/peabody-museum-discovers-possible-slave-remains-in-its-collections?page=all.

Bouffioux, M. (2020). Decolonization: The Free University of Brussels will return skulls to the Congolese. Retrieved from https://parismatch.be/actualites/societe/437106/decolonisation-luniversite-libre-de-bruxelles-va-restituer-des-cranes-aux-congolais.

Buckley, R., Morris, M., Appleby, J., King, T., O'Sullivan, D., & Foxhall, L. (2013). 'The king in the car park': New light on the death and burial of Richard III in the Grey Friars church, Leicester, in 1485. *Antiquity*, *87*(336), 519–538. https://doi.org/10.1017/s0003598x00049103.

Cloete, K., & Shaw, G. (2019). Restoring ancestors to their home through transformational UCT process [Press release]. Retrieved from https://www.news.uct.ac.za/features/sutherland/-article/2019-11-04-restoring-ancestors-to-their-home-through-transformational-uct-process.

Crais, C. C., & Scully, P. (2009). *Sara Baartmann and the Hottentot Venus: A ghost story and a biography*. Princeton University Press.

Dunnavant, J., Justinvil, D., & Colwell, C. (2021). Craft an African American graves protection and repatriation act. *Nature*, (7859), 337–344. https://doi.org/10.1038/d41586-021-01320-4.

Fauvelle-Aymar, F. X. (2002). *L'Invention du Hottentot: Histoire du regard occidental Sur les Khoisan (XVe-XIXe siécle)*. Paris: Publications du Sorbonne.

Freiburg, U. o. (2014). Repatriation of Skulls from Namibia [Press release]. Retrieved from https://web.archive.org/web/20150423033450/http:/www.pr.uni-freiburg.de/pm/2014/pm.2014-03-04.18-en.

Friedling, L. J., & Morris, A. G. (2005). The frequency of culture derived dental modification practices on the Cape Flats in the Western Cape. *South African Dental Journal*, *97*, 99–102.

Goose, D. H. (1963). Tooth mutilations in West Africans. *Man, 63*, 91–93.
Hochschild, A. (1998). *King Leopold's Ghost: A story of greed, terror, and heroism in colonial Africa.* Mariner Books.
Legassick, M., & Rassool, C. (2000). *Skeletons in the Cupboard: South African museums and the trade in human remains 1907–1917.* Cape Town: South African Museum.
Leslie, M., & Maggs, T. (Eds.). (2000). *Vol. 8. African Naissance: The Limpopo Valley 1000 years ago* South African Archaeological Society.
Mitchell, P. (2002). *The archaeology of Southern Africa.* Cambridge, UK: Cambridge University Press.
M-Net. (2019). Carte Blanche [Television series episode]. In M-Net (Executive producer) (Ed.), *The Way Back Home: Returning the Bones.* December 4, 2019.
Morris, A. G. (2011). *Missing & murdered: A personal adventure in forensic anthropology.* Cape Town, SA: Zebra Press.
Nienaber, W. C., Keough, N., Steyn, M., & Meiring, J. H. (2008). Reburial of the Mapungubwe human remains: An overview of process and procedure. *South African Archaeological Bulletin, 63*(188), 164–178.
Pakenham, T. (1991). *The scramble for Africa: The white man's conquest of the dark continent from 1867 to 1912.* Random House.
Ramsay, J. (2019). The humiliating and haunting return of El Negro to Africa. *Mmegi.* [The Reporter] Online, March 22, 2019. Retrieved from https://www.mmegi.bw/index.php?aid=80168&dir=2019/march/22.
Rassool, C. (2015). Human remains, the disciplines of the dead, and the south African memorial complex. In D. R. Peterson, K. Gavua, & C. Rassool (Eds.), *The politics of heritage in Africa: Economies, histories and infrastructures* (pp. 133–156). New York: Cambridge University Press.
Renschler, E. S., & Monge, J. M. (2013). The crania of African origin in the Samuel G. Morton cranial collection. *South African Archaeological Society Goodwin Series,* 35–38.
Schroeder, D., Chennells, R., Louw, C., Snyders, L., & Hodges, T. (2020). The rooibos benefit sharing agreement-breaking new ground with respect, honesty, fairness, and care. *Cambridge Quarterly of Healthcare Ethics, 29*(2), 285–301. https://doi.org/10.1017/s0963180119001075.
Stanton, W. R. (1960). *The Leopard's spots: Scientific attitudes towards race in America, 1915–1859.* Chicago, Illinois: University of Chicago Press.
Steyn, M., Morris, A. G., Mosothwane, M. N., Nienaber, W. C., & Maat, G. J. R. (2013). Introduction: Opening the cupboard—lessons in biology and history from African skeletons. *South African Archaeological Society Goodwin Series, 11*, 1–6.
Swingler, H. (2019). *'We knew their names' [Press release].* Retrieved from https://www.news.uct.ac.za/article/-2019-11-04-we-knew-their-names.
Tiley-Nel, S. (2016). Meet the 800-year-old golden rhinoceros that challenged apartheid South Africa. In *The conversation.* Retrieved from https://theconversation.com/meet-the-800-year-old-golden-rhinoceros-that-challenged-apartheid-south-africa-64093.
Wang, K., Goldstein, S., Bleasdale, M., Clist, B., Bostoen, K., Bakwa-Lufu, P., … Schiffels, S. (2020). Ancient genomes reveal complex patterns of population movement, interaction, and replacement in sub-Saharan Africa. *Science Advances, 6*(24). https://doi.org/10.1126/sciadv.aaz0183. eaaz0183.
Westerman, F. (2019). *El negro and me (revised).* Amsterdam: Atlas.
Wild, S. (2019). Restoring dignity to stolen ancestors. In *Sapiens.* 5 Dec 2019, Retrieved from https://www.sapiens.org/archaeology/south-africa-repatriation/.
Zuma, J. (2016). *Speech on the occasion of the reburial of Mr. Klaas and Mrs. Trooi Pienaar at Kuruman, Northern Cape Province, 12 August 2012.* Retrieved from http://www.northern-cape.gov.za/index.php/news-room/news-and-speeches/152-media-room/office-of-the-premier/news-and-speeches/769-reburial-of-mr-and-mrs-klaas-and-trooi-pienaar.

CHAPTER 13

Bones and objects of contention

Prologue

The 21st century promises to be a time of tension between global and national forces. News reports give daily attention to political and economic examples of this tension. Will arguments for open or closed borders prevail? Will the cultural norms of certain nations spread across boundaries, generating a global citizenry and possibly saving the planet's climate? Initiatives undertaken to press for return of plundered artworks and ancestral skeletons fall into this same tumult-filled basket of ideas. The argument that the British Museum should return the "Elgin Marbles" to Greece achieved a new level of public awareness when the issue was tabled during Brexit discussions. There are parallels between the arguments supporting quests for return of objects and quests for the return of ancestral bones. While these goals can often make common cause, there are also some important differences.

Repatriation refers to the transfer from one nation (*patria*, Latin for "one's own country") to another, so on the international scene this is often the right word. The goal is to send something or someone back to their own country. Sometimes the geographical context is not the best framework for the discussion. The issues may be within a country, or the group seeking a remedy may be dispersed across more than one country. For such contexts, the term restitution is often used. Restitution, in this context, means the restoration to its proper owner of something lost or stolen. While it is easy to shift repatriation to its verb form—repatriate—it is less graceful or familiar to talk of restituting. Merriam-Webster assures us that it is, nevertheless, a verb. Transitive.

In discussions of both repatriating and restituting, arguments about rightful ownership, respect, public access and appreciation of the object(s), responsible curation and use by future generations are weighed. Arguments about repatriating objects often center on the ability of the receiving country to provide suitable, secure permanent curation. When the bones of past humans are at issue, the topics of rightful ownership and respect arise, but

Osteobiographies
https://doi.org/10.1016/B978-0-12-823880-6.00021-5

the intention of the receiving group to remove the remains from curation—typically by burying them—creates a different context. It is not long-term management, but rather short-term alleviation of grievance that drives discussion forward. In the case of objects, assurances can be sought that access and scholarship on them will continue. In the case of human remains, consideration of past and future scholarship is often brief, or absent. There is little or no public pressure on institutions to collaborate with claimants to make sure the bones' value to scholarship is realized before repatriation and burial occur. Nevertheless, global support for cultural restitution affects attitudes and actions relating to human remains.

The world of objects

Perhaps the most well-known success for the argument supporting return of curated cultural items is the return to descendants of art that had been confiscated by the Nazis during World War II. Adolf Hitler had made detailed plans for a Führermuseum, to hold thousands of artworks plundered from (mostly) Jewish homes. The challenges faced by descendants seeking return of their property are dramatized in the 2015 film *Woman in Gold*. The film dramatizes the return of Klimt's *Portrait of Adele Block-Bauer* by the Austrian government to a descendant of the owner—although the transfer was neither quick nor easy. While the number of items that have made their way back to descendants is small, these acts are of symbolic importance. Governments and international professional associations have urged scrutiny of institutional accessions dating from 1933 to 1945 for signs of Nazi taint. There is broad acceptance of the virtue of repatriation in these cases.

This and other examples of physical property repatriation are encouraged by the 1970 UNESCO Convention on the Means of Prohibiting and Preventing the Illicit Import, Export and Transport of Ownership of Cultural Property (hereafter the UNESCO Convention). Its aspirations are articulated by an Intergovernmental Committee with a similarly long name, and a complementary Convention from UNIDROIT on the private law relevant to Stolen or Illegally Exported Cultural Objects. Another UN body, since 1946, is the International Council of Museums (ICOM). The Code of Ethics of ICOM emphasizes the need for valid title and says that "(6.2) Museums should be prepared to initiate dialogue for the return of cultural property to a country or people of origin. This should be undertaken in an impartial manner, based on scientific, professional and humanitarian principles as well as applicable local, national and international legislation, in preference to action at a governmental or political level" (Museums, 2017). The Code is at least permissive with respect to deaccessioning, which is a necessary preamble to repatriation activities. The UNESCO Convention has been ratified by 140 nations. Its actual impact is limited. Not all nations have signed on, and it is not retroactive. Because it is not retroactive, it does not address the booty in European museums from the scramble for objects and bodies that typified the colonial era, between 1885 and 1960. It also lacks any system of enforcement. While they do not have the weight of laws, these international initiatives have had some steering effect on public attitudes and they are said to have limited the intensity of the black market in antiquities (Robertson, 2019). They do not specifically address the management of human remains.

Unsurprisingly, not everyone is a fan of the idea of repatriation. About twenty years ago, nineteen museums in North America and Europe issued "The Declaration on the Importance and Value of Universal Museums" (Anonymous, 2004). Also referred to as encyclopedic/enlightenment/cosmopolitan museums, they aspire to represent the history of the entire world. As such, they argue that "museums serve not just the citizens of one nation but the people of every nation." The deaccessioning associated with repatriation would be a "disservice to all visitors."

Challenges to the validity of this assertion of special status have focused on objects of cultural patrimony held by these museums, like the Benin bronzes (Hicks, 2020; Marshall, 2020), the "Elgin" marbles (Robertson, 2019) and Easter Island moai (Zakaria, 2018). A 2018 report prepared for French president Emmanuel Macron estimated that 90% of African art and antiquities currently resides outside of the African continent. Most of the objects were taken during the colonial period. The report estimates that the British Museum holds 69,000 of these objects (Sarr & Savoy, 2018). This document, now called "the Macron report," advocates for return of African objects, and Macron pledged to act on its recommendations "without delay." The topic of cultural restitution through the return of objects to their geographic origins generates lively discussion. In addition to the ideological challenge presented by the universal museum argument, there are practical challenges, since political boundaries and cultural priorities have often changed substantially since the eighteenhundreds. Nevertheless, arguments for return(s) are gathering momentum. An example is the wittily titled recent book, "The Brutish Museums" (Hicks, 2020). The title of a review in *The Guardian* summarizes that author's message: "Return Everything" (Riley, 2020). The rationale is that repatriation will allow the descendants of those who created the objects to appreciate and build on their heritage.

The world of bones

What does all this have to do with the repatriation of human remains? Quite a bit. Like objects, human bones are accessioned within government-sanctioned institutions. Like objects, many were obtained during the colonial period, reflecting ideologies that seem quite foreign today. Like objects, each case presents unique ideological and practical challenges.

Curated collections of human remains can be broadly considered as arising from three different origins, although there is considerable overlap. The legacy collections come from the era of race science. They have been the focus of much of this book. They and the collections that arise from archeological excavations and chance discoveries (which I will call archaeology-origin) are my focus here. Some of the units within a legacy collection may have come from legitimate archeological excavations (as distinct from grave robbing). This illustrates the overlap between "bins." Collections comprised of the skeletons of cadavers that were and are obtained by medical programs through government-sanctioned means are a third category. The maintenance of what I am calling anatomy-origin collections is foundational to certain areas of teaching and research. Their maintenance and use are narrowly delimited by regional laws and regulations (Fig. 13.1). Their histories and modern contexts can be viewed as problematic (see for example de la Cova, 2019; Rassool, 2015), illustrating again that there is overlap among the contexts in which human remains are curated (Box 13.1). Still, they have not been the focus of repatriation efforts, while legacy and archaeology-origin collections have.

Most human remains in legacy collections are solely crania (without post-cranial skeletons), usually of adults. In the days of race science, it was believed that the skull was the most important source of scientific information about group identity, and classification of groups into races was the main purpose of research. Associated with racial classification came phrenology: the study of bumps on the skull as a tool for determining character and mental abilities. Some institutions combine the curation of legacy and archaeology-origin holdings as single collections.

The catalogued units of archaeology-origin collections come from archeological sites that range in age from recent to thousands of years ago. Most archaeology-origin collections are associated with excavations the purposes of which were to elucidate past human societies, including their burial practices. Some are fossils, meaning that the original bone tissue has been replaced by mineral from the surrounding soil. Within some museum collections, the antiquity of fossilized human remains may exceed that of

Fig. 13.1 Storage at an anatomy collection of human skeletons. *(Credit: S. Pfeiffer.)*

BOX 13.1 Anatomical collections

Body donation—the opportunity to donate one's body for medical training purposes—is a global feature of biomedical education. However, even today unclaimed bodies remain a major source of cadavers in dissection rooms across the globe, as they have been since the Late Middle Ages (Habicht, Kiessling, & Winkelmann, 2018). Throughout the history of medical education, some cadavers have been "skeletonized," so that study of the bones can provide ongoing educational benefits. Anatomical supply companies provided human skeletons for teaching, as well. Whether the bones are retained or not, a cadaver is a valued resource to which only selected learners have access. There is a strong trend to change the standard term from "cadaver" to "donor." Teaching programs have strict protocols that often include culturally specific commemorative ceremonies (Bolt, 2012).

There are skeletal collections associated with dissection programs on every continent. Most retain at least some information about the person in life (like gender, age at death, cause of death, body weight and so on). The extent to which the skeleton is "known" enhances its research value. These data allow assessment of the biases and limitations of each collection (Komar & Grivas, 2008). There are also forensically-oriented collections in which the natural decomposition of the donor's body is the focus of study (Christensen, 2006), and skeletal collections from decommissioned cemeteries. The latter are especially valuable for research, because they include the skeletons of known-age children.

Continued

BOX 13.1 Anatomical collections—cont'd

Many anatomical collections are impressively well curated. They are studied by qualified researchers who must comply with ethically and scientifically sensitive, detailed protocols. The collections have provided information about human health and diversity. Those studies contribute to our understanding of normal human growth, development and aging. Without them, forensic anthropologists could not supply informed, accurate estimates of the sex and ancestry of unknown persons. The collections have provided predictive equations for estimation of age-at-death, known-case comparisons for palaeopathological diagnoses, and samples of pathogenic aDNA. They are used to develop methods of study, like surface characterization (morphometrics) and three-dimensional imaging, which then can be applied to rare, fragmentary fossils. Some recent research has focused on what can be learned about the lives of the marginalized people represented by unclaimed bodies (Mant & Holland, 2019). Anatomical collections are important sources of information about the past lives of diverse human groups—their activities, their sizes and shapes, their aches and pains.

References

Bolt, S. (2012). Dead Bodies Matter: Gift Giving and the Unveiling of Body Donor Monuments in the Netherlands. *Medical Anthropology Quarterly, 26*(4), 613–634. https://doi.org/10.1111/maq.12010

Christensen, A. M. (2006). Moral considerations in body donation for scientific research: A unique look at the University of Tennessee's Anthropological Research Facility. *Bioethics, 20*(3), 136–145. https://doi.org/10.1111/j.1467-8519.2006.00487.x

Habicht, J. L., Kiessling, C., & Winkelmann, A. (2018). Bodies for Anatomy Education in Medical Schools: An Overview of the Sources of Cadavers Worldwide. *Academic Medicine, 93*(9), 1293–1300. https://doi.org/10.1097/acm.0000000000002227

Komar, D. A., & Grivas, C. (2008). Manufactured populations: What do contemporary reference skeletal collections represent? A comparative study using the Maxwell Museum documented collection. *American Journal of Physical Anthropology, 137*(2), 224–233. https://doi.org/10.1002/ajpa.20858

Mant, M. L., & Holland, A. J. (Eds.). (2019). *Bioarchaeology of Marginalized People*. New York: Elsevier.

objects by orders of magnitude. There are also collections of rather recently deceased humans that are of scholarly value because of their historical contexts as well as their genomes. For example, the curation of skeletons excavated from pits associated with the European Black Death made it possible to confirm *Yersinia pestis* as the cause of the disease (Bos et al., 2012).

Legacy collections

As is the case with objects, some accessioned units are associated with distressing histories. For example, William Henry Flower was curator at the British Natural History Museum (1884–1898). He and his predecessor Sir Richard Owen corresponded with government representatives in the far-flung British Empire of that era, requesting the procurement of human skeletons that would accurately represent the bodies of the natives: their physiques, facial features, and especially their skulls. Although normal practice was to send skeletons to London, skeletonizing of the deceased was not a requirement. An 1867 issue of the *Journal of Anatomy and Physiology* documents Flower's "Account of the Dissection of a Bushwoman." The body had been sent to him, in London. The practice of sending human remains continued well into the 20th century.

I visited the British Museum in 1996 to study the adult body size of skeletons from the Andaman Islands. The Islands make up an archipelago in the Bay of Bengal where the British established a penal colony in 1789. The indigenous tribes, unique in their small adult body size, used outrigger canoes to harvest marine resources (Stock & Migliano, 2009). Their skeletons are a valuable source of information about human diversity in body size and the effect of habitual behaviors on bone shape and mass. The Andaman Islanders had exceptionally strong shoulders and arms. I appreciated their availability, the careful curation of the collection for over a century, and the Museum's permission to study them. This did not lessen my uneasiness as I read accompanying labels like these:

- Andaman 1913; Chana Aga, wife of Bira, age 25? Female Andamanese, died of syphilis.
- Skeleton of Jarawa baby, captured when attacking a camp in Feb. 1902, died at Port Blair a few months later aged about 1 year. Skeleton should be complete, it was prepared by Dr. Sa..., assistant medical officer. [signed 1/9/05].
- From Little Andaman. Record of who he was and how he died in envelope with bones.

I make no claim to know about all the museums of the world, but those that I have visited usually have some curated units that make me uneasy because of the circumstances under which they were collected. Not all these institutions are in the lands of the colonizers. Museums established in the colonized lands often followed the same rubric. As noted in earlier chapters, the thin book titled *Skeletons in the Cupboard* (Legassick & Rassool, 2000) roused the curators of institutions in southern Africa to identify a category they call "unethically collected." They now prohibit research access to them, until they can figure out what to do about them. Anatomical collections and museums throughout the world are grappling with the challenges created by discussions of reparation and restitution as they are applied to human remains.

What is the scope of the human remains holdings in legacy collections? It is hard to absorb its magnitude, especially when considered in the context of these relatively small-scale repatriations. Simple on-line searches yield numbers like 19,000 specimens at the British Museum of Natural History, 18,000 individuals (most are crania) housed in the University of Cambridge's Duckworth Laboratory, 7000 at the Peabody Museum at Harvard University. Perhaps a more helpful approach to exploring the scope of the matter is to scan the distribution of the human remains from one previously colonized geographic region.

The indigenous peoples of southern Africa attracted the attention of race scientists from the northern hemisphere throughout the nineteenth century, and beyond. Through a grant from the South Africa Netherlands Partnership for Alternatives in Development (SANPAD), a modern team traveled broadly to identify southern African human remains held by institutions in Europe and the United States (Steyn, Morris, Mosothwane, Nienaber, & Maat, 2013). They focused on places associated with scholars who had special interests in sub-Saharan human anatomy. They did not attempt to include the many institutions that hold just one skeleton or a single skull. The common practice of anatomists exchanging "type specimens" would be very difficult to enumerate. An illustration: a few years ago, Alan Morris at University of Cape Town and I exchanged the "type" skeletons in the collections from our universities. I had seen in our records that my department held one "bushman." His department held one "red Indian." Archived correspondence confirmed the cordial exchange between former anatomists at our institutions—Professors Grant and Drennan—pursuing what was then seen as a valuable scholarly purpose. It was a kind of long-distance show-and-tell, to educate an international community of scholars about human racial diversity. While we had no details about their specific homes, we could at least get them back to their home continents, and we did.

As the SANPAD team scrutinized collections, they encountered ambiguities of provenance and labelling. Collections begun in the 19th century have been re-housed, amalgamated, mis-handled, broken and some collections were destroyed by bombing during World War II. The table below provides minimum estimates (Table 13.1). Although Khoe-San ancestors

Table 13.1 Sub-Saharan, especially Khoe-San, human remains curated at institutions outside Africa.

	Individuals represented
Natural History Museum, London	54
Duckworth Collection, Cambridge Univ.	11
Dept of Anatomy, Edinburgh Univ.	6
Anatomical Museum, Leids	19
Musée de l'Homme, Paris	15
Rudolf Pöch Collection, Vienna	147
Terry Collection, Smithsonian, D.C.	11
Peabody Museum, Harvard	5
American Museum of Natural History, NY	75
Morton Collection (crania), U Pennsylvania	84
Field Museum, Chicago	16

(Compiled from chapters in volume edited by Morris, A. G., (Ed.) (2013). Skeletal identity of past southern African populations: Lessons from outside South Africa. South African Archaeological Society, Goodwin Series, 11, 1–89.)

were the focus, in some cases the scope necessarily broadened to all indigenous sub-Saharan Africans. If there were ever a national initiative to repatriate South African indigenous ancestors, it would be an unwieldy project indeed.

There are ways in which human remains present challenges that go beyond those associated with objects. A request for repatriation may include both human remains and the objects that were funerary gifts or presumed possessions of the deceased. Such claims may differ from those that do not include objects. The consensus documents that support return of objects, like the UNESCO Convention, do not mention human remains. If bones and objects are to be considered together, the extension from cultural to biological heritage must be asserted. When the disposition of objects is discussed, the usual goal is to display them in, or near, their point of origin, where they were created. Questions are asked about security, climate control and curation. The expectation is that people will appreciate the objects for decades—if not centuries—to come. When the disposition of human remains is discussed, the end goal is very different.

Human bones are almost always in storage rather than on public display. The remains associated with a region or culture must be located through scrutiny of institutions' catalogues. Most frequently, the descendant group seeks to bury their ancestors' remains. In many cases, receiving communities request that there be little or no publicity, so only a select few are aware that the de-accessioning occurred. For some institutions, this can be frustrating, if they wish the public to know that they cooperated. Because arrangements for transfer of human remains may attract little public attention, there may be little motivation for curators to document and study the remains prior to de-accessioning. In other cases, the firm resistance to deaccessioning that has developed in response to the demand for objects can spill over into the discussion of human remains. Policy documents may constrain action (Box 13.2). The huge British Museum is bound by the British Museum Act (1963) to never permanently dispose of anything. Consistent with this mandate, the British Museum is a signatory to the Universal Museums declaration.

A clear, firm challenge to the keep-it-all stance came with a joint statement in 2000 from the British and Australian prime ministers, Tony Blair and John Howard, pledging to increase repatriation of ancestral remains to Australian indigenous communities. The British legal restriction against deaccessioning was addressed by passage of the 2004 Human Tissue Act. It includes a clause giving national museums the power to transfer human remains less than 1000 years old "if it appears to them to be appropriate to do so for any reason." This work-around legitimized several acts of repatriation, some broadly publicized and some quite low-key. Whether in the UK or elsewhere, each case presents its own unique conditions, created by the wishes and credentials of the claimants, and the policies of the curatorial institutions.

One of the first repatriations from Britain, in 2006, was the return of seventeen Tasmanians, to be cremated in accordance with local customs. This was in response to a request first made twenty years prior by the Tasmanian Aboriginal Centre (Anonymous, 2006). In 2011, ancestral bones were returned to Queensland, Australia. Parallel activities are underway across Europe. The skulls of 109 Maori and Moriori people were returned to New Zealand by the Charité hospital in Berlin in 2019 (Anonymous, 2019). New Zealand established and funded the Te Papa Museum in 2003, specifically to undertake a repatriation program to bring Maori ancestors home. They say they have repatriated hundreds of Maori and Moriori from overseas institutions. The French government created special legislation in 2011 to allow these actions. Negotiations with institutions elsewhere are said to be ongoing. The "Museums & Global Heritage" Facebook group has frequent notices of repatriation decisions.

BOX 13.2 Beothuk repatriation: An example of institutional challenges

In 2017, the skulls of two historically known Beothuk were returned to Canada. The saga behind this action illustrates the institutional challenges and the determination needed to achieve a return, whether of objects or human remains. Newfoundland is on the far eastern edge of Canada. It is known as "The Rock" for its challenging terrain and climate. The Beothuk lived in Newfoundland at the time of European arrival (Marshall, 1996); thereafter, they "faced the brunt of European commercial and religious activity" (Feit, 1999:26). The last known member of the community, the young woman Shawnadithit, died in 1829. The two crania are those of Nonosabasut and Demasduit. Demasduit was kidnapped by a European fur trapper in 1819. A watercolor of her, done while she was in captivity, is one of the few Beothuk images known (Fig. B13.1). Her husband Nonosabasut was killed when he resisted her capture. She is said to have died of "consumption" in 1820. Her death may have been hastened by trauma. The side of her skull shows

Fig. B13.1 The watercolor portrait of Demasduit, painted while she was held by the British in 1819. (https://en.wikipedia.org/wiki/Beothuk. Accessed Feb 3, 2021.)

Continued

BOX 13.2 Beothuk repatriation: An example of institutional challenges—cont'd

clear evidence of damage from a fall or a blow (Black, Marshall, & Kitchener, 2009). Europeans transported her coffin back to where she had been captured. In due course, she was placed in the burial hut of her husband. They were the aunt and uncle of Shawnadithit. It was from there that the skulls were collected in 1827. They were then sent to the University Museum in Edinburgh. The documentation is very thorough (Black et al., 2009; Pullman, 2018).

The Beothuk culture is considered extinct by government agencies. Nevertheless, there was broad interest in seeking the return of the remains of this couple. The request for return of the skulls was made by the Indigenous group that now lives on that land, the Mi'kmaq. The initial request was stymied by policies of the National Museum of Scotland in Edinburgh. Scottish legislation specified that remains may only be returned to direct descendants. So, leaders representing all Indigenous groups in the province of Newfoundland and Labrador signed a letter requesting the return of the remains, asserting thereby that they represent "a community descended from the original owners" (Pullman, 2018). This was acceptable. However, Scottish rules also indicated that the remains must be returned to a national institution. In 2019, after written confirmation of support from the Canadian federal government, the skulls were sent from Edinburgh (2000 miles east of Newfoundland) to the Canadian Museum of History (900 miles west of Newfoundland) (Shariatmadari, 2019). In March 2020, they were conveyed to Newfoundland, where a reburial is planned (CBC news, Mark Quinn, March 12, 2020).

References

Black, S. M., Marshall, I. C. L., & Kitchener, A. C. (2009). The Skulls of Chief Nonosabasut and His Wife Demasduit—Beothuk of Newfoundland. *International Journal of Osteoarchaeology, 19*(6), 659–677. https://doi.org/10.1002/oa.1004

Feit, H. A. (1999). Introduction: North America. In R. B. Lee & R. Daly (Eds.), *The Cambridge Encyclopedia of Hunters and Gatherers* (pp. 23–30). Cambridge, U.K.: Cambridge University Press.

Marshall, I. (1996). *A History and Ethnography of the Beothuk*. Montreal: Mc-Gill-Queen's University Press.

Pullman, D. (2018). Bioarchaeology, Bioethics, and the Beothuk. *American Anthropologist, 120*(1), 11–23. https://doi.org/10.1111/aman.12970

Shariatmadari, D. (2019). "They're not property': the people who want their ancestors back from British museums. *The Guardian*. Retrieved from https://www.theguardian.com/culture/2019/apr/23/theyre-not-property-the-people-who-want-their-ancestors-back-from-british-museums

Beyond museums

Discussions about the final disposition of human remains need to recognize that human remains—especially those that originate with archeological excavations—are quite often held by institutions that are not museums. These institutions include academic departments at universities, government agencies with heritage responsibilities, and independent archeological consulting companies. They are the main repositories for collections derived from archeological excavations. An example is the ancestral Huron-Wendat collections, discussed in Chapters 5 and 6. Often, accession and curation are not recognized parts of the institutions' mandate. Anywhere the profession of archaeology is taught or practiced, there is a strong possibility that excavated human remains are stored. Indeed, human remains are almost inevitably found in conjunction with cultural remains.

Wherever people lived, they may have left behind a few teeth—perhaps evulsed baby teeth, or the sources of toothaches that were pulled for relief. More substantively, people are normally buried or cremated near where they lived. Those archeological features may remain undisturbed until an excavation is undertaken, or they may be disturbed by burrowing animals, erosion, agricultural equipment, or trenches for any of the diverse subterranean installations of modern life. Discovered human remains can be minimal or massive in extent. They can be associated with clear markers of identity, or they can be anonymous. Once found, they often end up in boxes at some type of institution. The people responsible for holding these collections are not particularly expert in matters of curation, nor are they members of museum associations, with their codes of ethics and guides for best practices. They may feel pressures on matters of staffing and space and may operate without an oversight body. This was apparently the case at the Ethnographic Museum at a small Canadian university in the late 1990's. More than forty artifacts that were on loan from an Indigenous community went missing. Although most of them are still missing, some items were located years later in the United States and were returned, thanks to persistent investigation and negotiation (Matthews, 2016). It is easy to imagine that human remains would be at least as easy to remove and much harder to re-locate.

While the days of colonial administrators filling curators' orders for racially unique skulls are now long gone, the number of displaced human skeletons is likely to grow. Encounters with the remains of past generations are familiar features of civil engineering excavations. Annual field schools use historic cemeteries to train students in skeletal excavation methods. All these activities yield more skeletons. Those that are not re-buried in place will need to be catalogued and curated. They should also be studied.

The way forward

While far from exhaustive, these examples of objects and human remains that have changed hands provide a sense of the current landscape. The public understands that our museums hold huge collections of objects. The substantial extent and diversity of human remains under curation is less apparent to most people. The collections associated with anatomy programs (Box 13.1) were built and are managed under clear legal authority that is likely to preclude repatriation requests. Other collections, within the bins I have labeled "legacy" and "archaeology-origin," have varying amounts of latitude in their approach if a repatriation request is made.

When repatriation is requested, there are a lot of variables in play and each instance is unique. Not only can the holding institutions be diverse, so can the claimants. The simplest scenario may be when a claim comes from an autonomous (to some extent self-governed) descendant group whose members live more or less where their ancestors lived and who all agree on a path of action. The simplest timeline is relatively brief (say, ancestors from no more than a few hundred years ago) and knowledge of the cultural traditions has been maintained. This provides a clear cultural link between ancestor and descendant. Things can get more complicated in many ways: when the timeline is long, cultural histories are disparate, biological relatedness is less clear, the putative descendants are geographically dispersed, or there are competing claims. Complications are unlikely to dissuade contemporary people from seeking redress, respect and closure through the repatriation or restitution of human remains they perceive to be ancestral (Wade, 2021). Inevitably, some requests will be deigned to be specious. Who decides? How will decisions be made? Most importantly, can these challenges be framed as opportunities to expand everyone's knowledge and understanding?

It is unlikely that the transfer of human remains to descendant groups will ever be the topic of its own international consensus document. In light of this, I argue that the principles expressed in documents addressing return of cultural property can and should be extended to human remains. When a close relationship between a claimant and specifically identified human remains can be documented, the principles of the UNESCO Convention should apply. Following ICOM principles, curators—including curators-designate—should be willing to consider de-accession when a request is made. The established mandates of institutions will continue to present challenges, although there are now precedents for circumventions, as noted in this chapter. There is an analogy from the fields of genomics and palaeogenomics. Broadly accepted guidelines and protocols have been established for taking samples from living people, but not from the dead. It has been argued that there is value in thinking of ancient human remains as sources of biological information that require the same ethical consideration as research on living people (see Gibbon, 2020 for an example). Indeed, the value of a good materials transfer agreement that legally binds participating institutions is clear, whether the material is a Bronze Age urn, fresh human saliva, or a 200-year-old skull. Guidelines for how to tackle the challenge of permissions and transfers of authority can be borrowed and shaped, as long as the actors are willing.

This book has illustrated that there are many ways to achieve repatriation. The cases summarized in this and prior chapters have been selected to illustrate the diversity of approaches that are needed when the authority over human remains is transferred from curator to descendant. When the topic is considered more broadly, going well beyond the cases highlighted here, the feature that is most commonly missing from past and current repatriation dialogues is the engagement of scientists. When an institution is approached about de-accessioning human remains, that initiative should trigger outreach and dialogue that includes match-making between descendants and scientists.

Repatriation and restitution should include not only the transfer of authority for the remains of the ancestors, but also the creation of new knowledge about those ancestors, organized around questions that are of interest to the descendants (this sentiment is conveyed in an Australian context by Westaway, Clark, Williams, & Quayle, 2021). As noted in Chapter 5, the active involvement of people who know how to discover new things from old bones can help to dissolve cultural barriers and create new knowledge about questions that the descendants care about. Many institutions holding human remains do not have specialist scientists on staff. Such institutions could approach researchers who previously studied the collections in the context of their individual research programs. Those researchers should contribute their skills and knowledge to the repatriation as an aspect of ethical professional practice. As illustrated in previous chapters, a descriptive osteobiography can be enhanced with information from approaches that include dietary and environmental isotope values, facial reconstruction, radiocarbon dating as well as a palaeogenomic picture of the individual and their diseases. Research goals should be collaboratively determined with the descendant community. The proper, thorough documentation of human remains should be as clearly necessary as the thorough documentation of an object would be. With mindful, thorough handling, repatriations of human remains from legacy and archaeology-origin collections can be valuable examples of restitution and empowerment, as well as being sources of new knowledge.

References

Anonymous. (2004). Declaration on the importance and value of universal Museums. *ICOM News, 1*, 1.

Anonymous. (2006). British museum to return Tas aboriginal remains. In *Reuters*. Retrieved from https://www.abc.net.au/news/2006-11-18/british-museum-to-return-tas-aboriginal-remains/1312296.

Anonymous. (2019). *Berlin museum returns Maori and Moriori remains to New Zealand. www.DW.COM*. Retrieved from dw.com/en/berlin-museum-returns-maori-and-moriori-remains-to-new-zealand/a-48536886.

Bos, K. I., Stevens, P., Nieselt, K., Poinar, H. N., DeWitte, S. N., & Krause, J. (2012). Yersinia pestis: New evidence for an old infection. *PLoS One*, 7(11). https://doi.org/10.1371/journal.pone.0049803.

de la Cova, C. (2019). Marginalized bodies and the construction of the Robert J. Terry anatomical skeletal collection: A promised land lost. In M. L. Mant, & A. Holland (Eds.), *Bioarchaeology of marginalized people* (pp. 133–155). Elsevier Science & Technology.

Gibbon, V. E. (2020). African ancient DNA research requires robust ethics and permission protocols. *Nature Reviews Genetics, 21*(11), 645–647. https://doi.org/10.1038/s41576-020-00285-w.

Hicks, D. (2020). *The Brutish Museums: The Benin Bronzes, colonial violence and cultural restitution*. Pluto Press.

Legassick, M., & Rassool, C. (2000). *Skeletons in the cupboard: South African museums and the trade in human remains 1907–1917*. Cape Town: South African Museum.

Marshall, A. (2020, January 26). Will these treasures ever go home? *New York Times*. p. 2.

Matthews, M. (2016). *Naamiwan's drum: The story of a contested repatriation of Anishinaabe artifacts*. Toronto, Ontario: University of Toronto Press.

Museums, I. C. O. (2017). *ICOM code of ethics for museums*. ICOM.

Rassool, C. (2015). Re-storing the skeletons of empire: Return, reburial and rehumanisation of Southern Africa. *Journal of Southern African Studies, 41*, 653–670.

Riley, C. L. (2020, November 6). The Brutish Museums by Dan Hicks review—Return Everything. In *The Guardian*.

Robertson, G. (2019). *Who owns history? Elgin's Loot and the Case for Returning Plundered Treasure*. Knopf Australia.

Sarr, F., & Savoy, B. (2018). *The restitution of African Cultural Heritage. Toward a new relational ethics*. Retrieved from France http://restitutionreport2018.com/sarr_savoy_en.pdf.

Steyn, M., Morris, A. G., Mosothwane, M. N., Nienaber, W. C., & Maat, G. J. R. (2013). Introduction: Opening the cupboard—Lessons in biology and history from African skeletons. *South African Archaeological Society Goodwin Series, 11*, 1–6.

Stock, J. T., & Migliano, A. B. (2009). Stature, mortality, and life history among indigenous populations of the Andaman Islands, 1871-1986. *Current Anthropology, 50*(5), 713–725. https://doi.org/10.1086/605429.

Wade, L. (2021). The ghosts in the museum. *Science*, *373*(6551), 148–152.
Westaway, M., Clark, W., Williams, D., & Quayle, G. (2021). Reburying Barkandji human remains would close window on their past. *Nature*, *589*(7 January), 19.
Zakaria, R. (2018). Looted art and the 'universal museum': Can 21st-century collections ever escape colonialism's violent legacy? In *Frieze*. Retrieved from https://frieze.com/article/looted-art-and-universal-museum-can-21st-century-collections-ever-escape-colonialisms.

Index

Note: Page numbers followed by *f* indicate figures, *t* indicate tables, and *b* indicate boxes.

A

Accidents, 154, 168
African Humid Period, 113
Age at death, 23, 25–29
Age estimation, 155
Aggression, 165
American Journal of Physical Anthropology, 9–10
Amputation, 31–32
Anatomical collections, 197*b*, 199–200
Anatomy, 10
Animal teeth, 120–122
Archaeology-origin collections, 196–198, 206
Archaic farmers, 41–45, 49
Autopsy, 21, 29

B

Bantu-speakers, 138
Baartman, Sara, 185
Beothuk repatriation, 203*b*
Bioarchaeology, 1–2, 45*b*
Bioarchaeology of care, 62
Biological anthropology, 2, 12–13
Blunt force trauma, 163*b*
Bone tissue, modeling and remodeling of, 58*b*
Boskop, 139–140
Breastfeeding, 26–29
British Museum Act (1963), 201–202
Burial archaeology, 45*b*

C

Calipers, 13
Canadian Association for Physical Anthropology (CAPA), 84–85
Cape Floristic Region, 130
Cape Fold Belt, 131
Cape Town, 125–132, 141
Carbon isotopes, 13, 74–79
Cemeteries, 25
Cemetery archaeology, 45*b*
Childbirth, 152–154
Child growth, 150–152
Chronic stress indicators, 150–151
Coalitionary violence, 172
Coffins, 21–25, 31, 33
Collaboration, 83–96
Collections
  anatomical, 197*b*, 199–200
  archaeology-origin, 196–198, 206
  legacy, 196, 199–202, 206
  unethically collected, 199–200
Collective osteobiographies, 74–79
Colonialism, 130, 138, 177
Cremation, 51–55
Curation, 85, 90, 96

D

Death studies, 45*b*
Demography, 70–71
Dentition, 10, 55–57, 77–78, 118–119, 141
Diagenesis, 116–117
Diet, 150–151
Dietary isotopes, 77–79
Differential diagnosis, 59–60, 62, 72
Display, 186–187
Doves, 165–166

E

El Negro of Banyoles, 186–187
Environmental isotopes, 79
Ethics guidelines, 85
Ethnography, 16–17

F

Fairty Site Ossuary, 70, 70*f*
Farmworkers, 187–188
Feast of the Dead, 43, 44*f*
Feud, 172
First Nations groups, 37–38
Foraging, 149–150, 154–156
Forensic anthropology, 45*b*
Forensic archaeology, 45*b*

Fossils, 196–198
Founding First Nations Circle, 87–88
Funerary archaeology, 45*b*
Fynbos, 130–131, 133–141

G

Genomics, 101–102
Glacial Kame Burial Complex, 51–53

H

Habitual postures, 148
Hawks, 165
Healed cranial trauma, 166–168
Herding, 113, 122
Hind Site, 55–56, 61
History, 2, 18–19
Hobbesians, 165
Human bioarchaeology, 1
Human Tissue Act (2004), 202
Hunter-gatherers, 51–53, 62
Huron-Wendat Nation, 37–41, 43–45, 47–48, 65–67, 69, 71–76, 83–96

I

Illness, 154
Indigenous groups, 40–45, 47–49
In situ hypothesis, 67
International Council of Museums (ICOM), 194–195
Interpersonal violence, 170, 171*t*
Iron Age farmers, 138, 140–141
Ivory, 115, 120–123

J

Jesuit Relations, 65–66, 71

K

Kenya, 111–113, 114*f*
Khoe-San, 102, 109–110, 125–142, 146–155
Khoikhoi, 137–139
Kleinburg Ossuary, 18–19

L

Lake Turkana, 112–124
Late Archaic, 51–53
Later Prehistory of West Turkana unit, 111
Later Stone Age, 133–137, 134*b*, 135*f*, 145
Legacy collections, 177–178, 183, 196, 199–202, 206
Life expectancy, 154–155
Logistic mobility, 54
Lower Great Lakes, 37–49

M

Macron report, 195
Manemanya, 117–123
Mapungubwe, 181–182
Material transfer agreement, 95–96
Middle Stone Age, 131–132
Migration hypothesis, 66–67, 69–70
Moatfield Ossuary, 68
Modern *Homo sapiens*, 101, 109
Molars, 93
Mortuary archaeology, 45*b*
Mycobacterium tuberculosis, 15–16, 69–70, 72–73, 154

N

Nama and Herero genocide, 179–180
Native American Graves Protection and Repatriation Act (NAGPRA), 7–9
Nitrogen isotopes, 76–79
Northern Iroquoian-speaking groups, 38, 39*f*, 67, 72–73

O

Ontario Cemeteries Act, 21–23
Ontario Heritage Act, 46–47, 84–85
The Origin of Races, 102, 103*f*
Ossossané Ossuary, 18–19, 48, 65–66, 68–69, 84–85
Ossuary, 43–45, 47–48, 65–71, 74–79
Osteoarchaeology, 45*b*
Osteobiographies, 2–9
  Eastern and Southern Africa, 104–110
  Lake Turkana, 112–124
  Lower Great Lakes, 37–49
Osteogenesis imperfecta (OI), 60–61
Osteological paradox, 5–6
Ostrich eggshell, 117, 120–123

P

Palaeogenomics, 15–16, 105, 105*b*, 109–110
Palaeopathology, 72
Paleodemography, 72–73

Parachute science, 161
Pelvic dimensions, 152–154
Personal identification, 35
Phrenology, 196
Physical anthropology, 10
Pienaar, Klaas and Trooi, 187–188
Pillar sites, 114–122
Poison, 166, 170
Policy on Indigenous Repatriation, 7
Population structure, 109
Prospect Hill, 21–30

R

Race science, 2–3, 7–8, 12–13, 68, 177, 182–183, 190
Racial classification, 102–104
Radiocarbon dating, 14*b*
Reburial, 37, 47–48
Red ochre, 54–57
Repatriation, 7–8, 83–96, 177–191, 193
  events, 7, 33, 83–84, 185
  laws, 7–9, 14
  policy, 7, 85, 89, 179
Research program, 99
Residential mobility, 54
Restitution, 177, 193
Rousseauians, 165–166
Royal Museum for Central Africa, 178

S

Schmorl's depression, 33, 34*f*
Scramble for Africa, 177, 178*f*
Slavery, 182, 185
Snake Hill, 30–35
South Africa Netherlands Partnership for Alternatives in Development (SANPAD), 200
Southern African foragers
  geographical and historical context, 131–142
  interpersonal violence, 161–162, 163*b*, 165–174
  past communities, life histories of, 145–156
Southern African Naissance, 181–182
Stable isotopes, 13–15
Stature, 132, 145–148, 156
Stone beads, 120–122
Strontium isotopes, 74–76, 79

T

Terminology, 92*t*
Thonnakona Ossuary, 94–95
Tooth formation, 26, 27*b*
Tooth wear, 58–59
Transfer of authority, 179
Tuberculosis, 69–70, 72–74
Turkana Basin, 112–114
Turkana Basin Institute (TBI), 111

U

UNESCO Convention, 194–195, 201, 206–207
Unethical collection, 184, 188–191, 199–200
Unhealed cranial trauma, 168–170
Universal museums, 195
University of Brussels (ULB), 179
University of Cape Town (UCT), 189
University of Lubumbashi (UNILU), 179
University of Toronto
  Anthropology Department Repatriation Policy, 85
U.S. Army Casualty and Mortuary Affairs Operations Division, 30–31

V

Vermillion Accord, 6–7
Victims, 168–172

W

Walpole Island First Nation, 61
War of 1812, 30, 32*f*
Weaning, 26–29
Wendake, 84–85, 87–88, 92–94
Woman in Gold, 194
Woodland farmers, 41–43

Kolibri 360 GmbH
Ferdinand-Jühlke-Straße 7
99095 Erfurt, Deutschland
produktsicherheit@kolibri360.de